THE DIY OFF GRID SOLAR POWER BIBLE

[10 IN 1]

The Most Complete and Updated Guide to Design,
Install and Maintain Solar Energy Systems for
Tiny Homes, Cabins, RVs, and Boats

JACKSON MITCHELL

TABLE OF CONTENTS

Introduction *1*

BOOK 1: INTRODUCTION TO OFF-GRID SOLAR POWER 3

Chapter 1: Understanding Off-Grid Solar Power Systems *4*

Grid-Tied vs. Off-Grid Solar Systems: Key Differences 4

Pros and Cons of Off-Grid Living with Solar Power 5

    Pros: 5

    Cons: 6

Real-Life Off-Grid Solar Power Success Stories 6

    Solar Power Empowers Himalayan Villages 6

    Island Sustainability in American Samoa 7

    Educational Advancement in Rwanda 7

    Off-Grid Resilience in Australia 7

Chapter 2: Advantages and Benefits of Going Off-Grid *8*

Energy Independence and Environmental Impact 8

Cost Savings and Return on Investment (ROI) 9

Off-Grid Living Lifestyle: Freedom and Sustainability 9

Chapter 3: Essential Components of a Solar Energy System *11*

Solar Panels: Types, Efficiency, and Wattage Calculation 12

    Types of Solar Panels 12

    Efficiency of Solar Panels 12

    Wattage Calculation 12

Deep-Cycle Batteries: Lead-Acid vs. Lithium-Ion 13

    Lead-Acid Batteries 13

    Lithium-Ion Batteries 13

    Choosing the Right Battery 14

Charge Controllers: PWM vs. MPPT and Proper Sizing..14

    PWM Charge Controllers..14

    MPPT Charge Controllers..14

    Proper Sizing of Charge Controllers 14

Chapter 4: Calculating Your Power Needs and Energy Consumption *16*

Energy Audit: Assessing Your Electricity Usage 16

    Gathering Data..17

    Identifying Energy Consumption Patterns..17

    Peak Usage Analysis..17

    Equipment and Appliance Assessment..17

    Building Envelope Evaluation 17

Estimating Daily Energy Requirements for Different Applications.......................................18

    Lighting the Way 18

    Powering Essential Appliances 18

    Staying Warm 18

    Charging Electronics..19

    Appliances with Variable Usage 19

Sizing Your Solar System to Meet Energy Demand 19

BOOK 2: SOLAR POWER SYSTEM DESIGN AND PLANNING 21

Chapter 1: Site Assessment and Solar Resource Analysis *22*

Conducting a Site Survey: Sunlight Exposure and Obstructions..23

    Sunlight Exposure Analysis 23

    Obstruction Identification...23

    Roof Analysis (For Rooftop Installations)..23

Analyzing Local Climate Data for Solar Energy Estimation 24

    Solar Irradiance Data..24

    Temperature and Heat Effects 24

    Weather Patterns 25

Choosing the Optimal Location for Solar Panel Installation 25

    Maximizing Sunlight Exposure...25

Considering Aesthetics and Regulations 25

Roof vs. Ground Installation 26

Structural Considerations..26

Long-Term Goals 26

Chapter 2: Sizing Your Solar System for Tiny Homes **28**

Understanding Load Profiles for Small Living Spaces 29

Off-Grid Solar Design Considerations for Tiny Homes.....................................29

Balancing Power Generation and Energy Consumption 30

Chapter 3: Designing Solar Systems for Cabins and Remote Locations **32**

Off-Grid Cabin Energy Needs: Seasonal Variations 33

Remote Location Challenges: Transporting and Installing Equipment 33

Backup Power Solutions for Extended Periods of Low Sunlight 34

Chapter 4: RV and Boat Solar System Design Considerations **36**

RV Solar Panels: Portable vs. Roof-Mounted Installations..................................37

Portable Solar Panels: Unleashing Flexibility...37

Roof-Mounted Solar Panels: Seamlessness and Efficiency...............................37

Marine Solar Systems: Waterproofing and Corrosion Protection 38

Waterproofing: Defying the Ingress of Moisture 38

Corrosion Protection: Battling the Salt and Humidity 38

Mobile Solar Power Solutions for Traveling and Boating 39

Battery Technology: Enabling Energy Storage..39

Portable Solar Panels: Unfolding Freedom 39

Integration and Hybrid Solutions: The Best of Both Worlds.............................40

BOOK 3: SELECTING THE RIGHT SOLAR EQUIPMENT 41

Chapter 1: Choosing the Best Solar Panels for Your Setup **42**

Monocrystalline, Polycrystalline, and Thin-Film Panels Comparison........................43

Monocrystalline Panels...43

Polycrystalline Panels..43

Thin-Film Panels...43

Evaluating Solar Panel Efficiency and Performance Ratings 44

    Temperature Coefficient 44

    Power Tolerance 44

Factors Affecting Solar Panel Lifespan and Maintenance 44

    Quality of Materials 44

    Weather Conditions 45

    Regular Maintenance 45

    Inverter Maintenance 45

    Shading and Placement 45

    Warranty and Support 45

Chapter 2: Evaluating Battery Options for Energy Storage 47

Lead-Acid Batteries: Deep-Cycle vs. Flooded vs. AGM 48

    Deep-Cycle Lead-Acid Batteries 48

    Flooded Lead-Acid Batteries 48

    AGM (Absorbent Glass Mat) Lead-Acid Batteries 48

    Considerations for Lead-Acid Batteries 48

Lithium-Ion Batteries: Benefits, Types, and Considerations 49

    Benefits of Lithium-Ion Batteries 49

    Types of Lithium-Ion Batteries 49

    Considerations for Lithium-Ion Batteries 50

Battery Capacity Sizing and System Voltage Configurations 50

    Battery Capacity Sizing 50

    System Voltage Configurations 51

Chapter 3: Inverters and Charge Controllers: Types and Selection 52

Off-Grid Inverters: Pure Sine Wave vs. Modified Sine Wave 53

    Pure Sine Wave Inverters 53

    Modified Sine Wave Inverters 53

MPPT Charge Controllers: Advantages and Optimal Use Cases 53

    Advantages 54

    Optimal Use Cases 54

Sizing Inverters and Charge Controllers for Efficient Energy Conversion 54

    Inverter Sizing 55

    Charge Controller Sizing 55

Chapter 4: Other Components: Cables, Mounting Systems, and Monitoring Devices 56

Solar Cables and Connectors: Proper Sizing and Protection 57

    Sizing Solar Cables 57

    Choosing Connectors 57

    Cable Length and Voltage Drop 58

    Protection Against the Elements 58

Mounting Solar Panels: Roof, Ground, Pole, and Tracking Systems 58

    Roof-Mounted Solar Panels 58

    Ground-Mounted Solar Arrays 59

    Pole-Mounted Solar Arrays 59

    Tracking Systems 59

    Balancing Efficiency and Practicality 59

Monitoring and Control Systems for Real-Time Performance Tracking 60

BOOK 4: INSTALLATION AND SETUP OF SOLAR SYSTEMS 61

Chapter 1: Safety Precautions and Installation Guidelines 62

Electrical Safety Measures: Working with Solar Components 63

    Grounding and Wiring Integrity 63

    DC Voltage Awareness 63

    Disconnect Procedures 64

Roof and Ground Installation Safety Practices 64

    Rooftop Installations 64

    Ground Installations 65

Adhering to Local Building Codes and Permit Requirements 65

    Navigating Building Codes 65

    Obtaining Permits 65

Chapter 2: Installing Solar Panels on Different Surfaces (Roofs, Ground, etc.) 67

Roof-Mounted Solar Panel Installation: Pitched and Flat Roofs 68

Pitched Roofs 68

Flat Roofs: 69

Ground-Mounted Solar Arrays: Foundation and Tilt Angles 70

Site Selection and Preparation 70

Mounting Structure and Tilt Angle Optimization: 70

Wiring, Inverter Setup, and Maintenance: 70

Portable Solar Panels: Setting Up Solar Arrays Anywhere 71

Design and Portability 71

Deployment and Use 71

Versatility and Maintenance 71

Chapter 3: Connecting Batteries and Configuring Energy Storage 73

Battery Bank Configuration: Series vs. Parallel Connections 74

Series Connections 74

Parallel Connections 74

Combined Connections 74

Proper Wiring and Fusing for Battery Safety and Efficiency 75

Battery Maintenance and Capacity Testing 76

Maintenance Practices: 76

Capacity Testing: 76

Chapter 4: Wiring Your Solar System: Step-by-Step Guide 78

Electrical Circuit Design and Wiring Diagrams 78

Planning Your Electrical Circuit 79

Choosing the Right Wire Sizes 79

Creating Wiring Diagrams 79

Differentiating AC and DC Wiring 79

Color Coding and Labeling 79

Wiring Solar Panels to Charge Controllers and Batteries 80

Connecting Solar Panels to Charge Controllers 80

Wiring Batteries to Charge Controllers 81

AC and DC Wiring for Power Distribution in Off-Grid Systems 81

BOOK 5: OFF-GRID POWER MANAGEMENT AND OPTIMIZATION 83

Chapter 1: Understanding Energy Load Management 84

Energy Load Analysis: Identifying High-Consumption Devices 85

Load Shifting and Energy Conservation Strategies 85

Load Priority and Managing Essential vs. Non-Essential Loads 86

Chapter 2: Energy Efficiency Tips for Tiny Living Spaces 87

Energy-Efficient Appliances and Lighting Solutions 87

Insulation and Passive Heating/Cooling Techniques 89

Smart Home Automation for Optimal Energy Usage 90

Chapter 3: Optimizing Power Consumption in Cabins and RVs 92

Energy-Saving Cabin Design and Insulation 93

RV Energy Management: Off-Grid vs. On-Grid Campsites 94

Sustainable Water and Waste Management Solutions 94

Chapter 4: Monitoring and Troubleshooting Your Off-Grid Solar System 96

Remote Monitoring and Data Analysis Tools 97

The Power of Remote Monitoring 97

Data Analysis: A Key to Optimization 97

Role of Remote Monitoring Tools 97

Diagnosing Common Solar System Issues and Failures 98

Identifying and Resolving Issues 98

Maintenance Best Practices to Prolong System Lifespan 99

BOOK 6: OFF-GRID SOLAR POWER FOR TINY HOMES 100

Chapter 1: Solar-Powered Heating and Cooling Solutions 101

Passive Solar Design: Harnessing Sunlight for Heating 102

Solar Space Heaters and Radiant Floor Heating 102

Solar Air Conditioners and Ventilation Systems 103

Chapter 2: Solar Water Heating Systems for Tiny Homes **104**

   Types of Solar Water Heaters: Batch, Flat-Plate, and Evacuated Tube 104

      Batch Solar Water Heater 105

      Flat-Plate Solar Water Heater 105

      Evacuated Tube Solar Water Heater 105

   DIY Solar Water Heater Construction and Installation 105

   Solar Water Pumps and Circulation in Off-Grid Systems 106

      Importance of Solar Water Pumps: 107

      How Solar Water Pumps Work: 107

Chapter 3: Solar-Powered Appliances and Gadgets **109**

   Solar Ovens and Cooking Appliances 110

   Solar-Powered Refrigerators and Freezers 110

   Solar-Powered Electronics and Charging Solutions 111

Chapter 4: Off-Grid Lighting: LED and Other Solutions **112**

   Energy-Efficient LED Lighting for Tiny Spaces 113

   Solar Lighting Systems: Indoor and Outdoor Applications 113

   Solar-Powered Pathway Lights and Security Lighting 114

BOOK 7: OFF-GRID SOLAR POWER FOR CABINS **115**

Chapter 1: Designing Solar-Powered Cabin Retreats **116**

   Off-Grid Cabin Design Considerations and Aesthetics 117

   Remote Cabin Power Supply: Determining Energy Needs 117

   Balancing Comfort and Sustainability in Cabin Living 118

Chapter 2: Cabin Off-Grid Plumbing Solutions **119**

   Rainwater Harvesting and Greywater Recycling 120

   Solar-Powered Water Pumps and Filtration Systems 121

   Composting Toilets and Waste Management 121

Chapter 3: Off-Grid Refrigeration and Food Storage **123**

   Propane vs. Solar-Powered Refrigeration: Weighing the Options 124

   Root Cellars and Passive Food Storage Techniques: Nurturing Nature's Wisdom 125

Solar Food Dehydrators and Preservation Methods: Modern Twists on Timeless Techniques 126

Chapter 4: Securing and Protecting Your Cabin Solar System 127

Cabin Security Systems and Surveillance 128

Security System Integration 128

Surveillance Cameras 128

Remote Monitoring 128

Alarm Systems 128

Lightning and Surge Protection for Solar Arrays 129

Guardians of Protection: Strategies and Measures 129

An Art Form of Installation and Vigilance 130

Winterizing Your Off-Grid Cabin and Solar Equipment 131

The Cocoon of Insulation and Sealing 131

Tilt and Snow Removal 131

Battery Care and Resilience 131

Antifreeze and Prudent Care 132

BOOK 8: OFF-GRID SOLAR POWER FOR RVS 133

Chapter 1: Installing Solar Panels on RVs and Motorhomes 134

Roof-Mounted vs. Portable Solar Panels: Pros and Cons 134

Roof-Mounted Solar Panels 135

Portable Solar Panels 135

Retrofitting RVs for Solar Power: Step-by-Step Guide 135

Maximizing Solar Energy Collection While Traveling 136

Chapter 2: Battery Upgrades and Energy Storage for Traveling 138

Lithium-Ion Battery Upgrades for RVs 138

Solar Generators and Portable Power Stations 139

RV Battery Maintenance and Longevity Tips 140

Chapter 3: Solar-Powered RV Appliances and Electronics 142

Solar-Powered RV Air Conditioners and Fans 142

Efficient RV Lighting and Entertainment Systems 143

Solar-Powered RV Water Heaters and Showers 143

Chapter 4: Solar RV Maintenance and Tips for Life on the Road *145*

Cleaning and Maintaining Solar Panels on the Go 145

Planning Your Routes Based on Solar Energy Availability 146

Staying Connected: Finding RV-Friendly Solar Campsites 147

BOOK 9: OFF-GRID SOLAR POWER FOR BOATS 148

Chapter 1: Marine Solar Panel Installation and Setup *149*

Boat-Specific Solar Panel Mounting Solutions 149

Fixed-Mount Systems 150

Adjustable Tilt Systems 150

Pole Mounts 150

Flexible and Lightweight Solar Panels for Boats 151

A Seafaring Evolution 151

Conquering Space Constraints 151

Navigating the Waters of Efficiency 151

Durability in the Face of Elements 151

The Aesthetic Symphony 152

Charting a Sustainable Course 152

Solar Panel Placement on Sailboats and Motor Yachts 152

Sailboats: 152

Motor Yachts: 153

Chapter 2: Battery Banks and Charging for Boats *154*

Marine Battery Bank Configuration and Placement 155

Battery Configuration 155

Placement Considerations 155

Solar Charge Controllers for Marine Environments 155

Role of Solar Charge Controllers 156

Types of Solar Charge Controllers 156

Integration Challenges 156

Wind and Hydro Turbines as Supplementary Charging Sources 156

Wind Turbines 157

Hydro Turbines 157

Integrating Turbines 157

Chapter 3: Solar-Powered Boat A ppliances and Navigation Systems **158**

Solar-Powered Boat Refrigeration and Cooking 158

Solar-Powered Refrigeration 159

Solar-Powered Cooking 159

Solar-Powered GPS and Communication Devices 160

Solar-Powered GPS 160

Solar-Powered Communication Devices 160

LED Navigation Lights and Solar Dock Lighting 161

Solar-Powered LED Navigation Lights 161

Solar Dock Lighting 161

Chapter 4: Off-Grid Living Afloat: Safety and Maintenance **163**

Marine Safety Equipment and Emergency Preparedness 164

Saltwater Corrosion Prevention and Rust Control 164

Maintaining Solar Panels and Equipment on Watercraft 165

BOOK 10: ADVANCED OFF-GRID SOLAR POWER CONCEPTS **167**

Chapter 1: Hybrid Solar Systems: Incorporating Other Energy Sources **168**

Integrating Wind Power with Solar for Increased Reliability 168

Hybrid Solar Generators: Biomass and Micro-Hydro 169

Grid-Interactive Systems and Net Metering 170

Chapter 2: Off-Grid Solar for Larger Applications and Homesteads **171**

Designing Off-Grid Solar for Farms and Homesteads 171

Large-Scale Energy Storage Solutions 172

Managing Power for Multi-Structure Off-Grid Systems 173

Chapter 3: Off-Grid Solar Entrepreneurship and Community Projects *175*

    Starting a Solar Installation Business 176

    Community-Based Solar Initiatives and Cooperative Projects 176

        Empowering the Collective 177

        The Cooperative Tapestry 177

        Dividends Beyond Energy 177

        Shared Costs, Shared Benefits 177

        A Blueprint for Sustainability 177

        Weaving a Sustainable Future 178

Chapter 4: Future Trends and Innovations in Off-Grid Solar Power *179*

    Advancements in Solar Panel Technology 179

    Energy Storage Breakthroughs and Emerging Battery Technologies 180

    The Evolution of Off-Grid Solar Power Systems 181

Conclusion *182*

INTRODUCTION

In an ever-evolving world, where our choices ripple beyond our own lives, a revolution in energy consumption is underway. As the twentieth century drew to a close, the narrative surrounding food and its impact on human health underwent a profound transformation. The focus shifted from merely preventing deficiencies and associated diseases to a higher goal: optimizing nutrition to enhance overall well-being and mitigate the risks of various illnesses. The pursuit of longevity was no longer the sole objective; the twenty-first century beckoned us to prioritize quality of life through what we consume. This paradigm shift marked the emergence of functional foods as a compelling concept, one rooted in scientific inquiry and poised to revolutionize dietary guidelines.

As the author of *The DIY Off-Grid Solar Power Bible*, I am an eminent figure in the world of do-it-yourself projects. My journey has been one of dedicated exploration, spanning various domains, from woodworking and metalworking to electronics and renewable energy systems. My passion for hands-on craftsmanship has ignited a lifelong commitment to sharing practical solutions and innovative ideas. Yet, beyond my skills, it's my unwavering dedication to sustainable living that has shaped my identity.

Within these pages, I am excited to introduce you to the intricate universe of off-grid solar power. As a trailblazer in the realm of sustainable living, I've harnessed the power of the sun to inspire countless individuals to embrace eco-friendly solutions. My mission is to empower you to design, construct, and manage your very own off-grid solar power systems. But this book isn't just about solar panels and batteries; it's about embarking on a journey of self-sufficiency, environmental consciousness, and empowerment.

The book isn't a mere manual; it's a comprehensive education encompassing a wide spectrum of topics, presented in a practical, reader-friendly manner. It's a guide that not only imparts knowledge but also cultivates a sense of capability—empowering even those new to the world of off-grid solar power to take charge of their energy destiny.

Through its ten distinct books, each devoted to a specific facet of off-grid solar power, this tome offers a meticulously structured curriculum. In Book 1, you'll explore the fundamental concepts of off-grid solar systems, differentiating them from their grid-tied counterparts and examining the advantages and potential challenges of embracing off-grid living. Real-life success stories provide inspiration and validation for your journey.

The subsequent books delve into every conceivable aspect of off-grid solar power, guided by three foundational pillars: comprehensive education, practical guidance, and adaptability. From understanding solar components and estimating power requirements to conducting site assessments and analyzing solar resources, each step is meticulously explored.

The heart of this opus lies in its ability to cater to diverse needs. Whether you're a tiny home enthusiast, a cabin dweller seeking refuge in nature, an intrepid traveler navigating the open road in an RV, or a soul who finds solace in the gentle lapping of waves on a boat, this book has a tailored offering for you. It's a testament to the versatility of off-grid solar power, seamlessly integrating into various lifestyles and aspirations.

Yet, this comprehensive manual goes beyond technical know-how. It explores the art of efficient energy management, optimizing consumption, and embracing sustainable practices in every corner of life. It is an ode to harmony between nature and technology, a guide to crafting a life that's not just environmentally friendly but rich in comfort and convenience.

As your guide on this odyssey, my approach is rooted in approachability. Just as my demeanor has earned me a dedicated following, I seek to be that friend who imparts knowledge with warmth and ease. I'm not just an author; I'm your companion in this journey, unraveling the intricacies of off-grid solar power by your side.

In a world where the choices we make reverberate far beyond our immediate surroundings, *The DIY Off-Grid Solar Power Bible* isn't just a book—it's a catalyst for change. It's about embracing a future where clean, renewable energy propels us forward. It's about saying yes to self-sufficiency, to sustainability, and to a world where every sunrise brings not just light, but the promise of power at our fingertips.

Join me in this transformative journey as we usher in a new era of energy, one illuminated by the brilliance of the sun and powered by the determination of individuals like you.

BOOK 1
INTRODUCTION TO OFF-GRID SOLAR POWER

CHAPTER 1: UNDERSTANDING OFF-GRID SOLAR POWER SYSTEMS

In an age where environmental consciousness and renewable energy sources are becoming paramount, off-grid solar power systems have garnered considerable attention. These systems are revolutionizing the way we generate and utilize energy, offering individuals and communities the opportunity to achieve energy self-sufficiency while reducing their carbon footprint. This chapter delves into the intricacies of off-grid solar power systems, elucidating the key differences between grid-tied and off-grid systems, examining the pros and cons of embracing off-grid living with solar power, and showcasing real-life success stories that underscore the potential and benefits of this innovative energy solution.

Grid-Tied vs. Off-Grid Solar Systems: Key Differences

The fundamental distinction between grid-tied and off-grid solar power systems lies in their interaction with the electricity grid. A grid-tied system is connected to the utility grid, allowing users to draw power from the grid when their solar panels do not produce enough energy and to feed excess energy back into the grid when their panels generate surplus power. This interaction enables users to balance their energy consumption and potentially receive credits for the excess energy they contribute.

On the other hand, off-grid solar power systems operate independently of the utility grid. These systems are designed to provide all the energy required for a household or establishment without relying on external sources. Off-grid setups consist of solar panels, energy storage systems (typically batteries), charge controllers, inverters, and backup generators. Solar panels capture sunlight and convert it into direct current (DC) electricity. Charge controllers regulate the flow of electricity to the batteries, preventing overcharging and prolonging their lifespan. Inverters convert DC electricity into alternating current (AC), which is used to power household appliances. Energy storage in batteries ensures a continuous power supply during cloudy days or nighttime.

Choosing a grid-tied or off-grid system depends on several factors, including geographical location, energy consumption, budget, and personal preferences. Grid-tied systems are simpler to install and often more cost-effective initially. They allow users to take advantage of net metering, reducing electricity bills by selling excess energy. However, these systems are vulnerable to grid failures, which means they won't provide power during blackouts or emergencies. In contrast, off-grid systems offer energy independence and resilience, ensuring a consistent power supply regardless of grid status. Yet, they require a larger upfront investment due to the need for energy storage solutions.

Pros and Cons of Off-Grid Living with Solar Power

Pros:

1. **Energy Independence:** One of the most appealing aspects of off-grid solar power systems is the liberation from the uncertainties of the grid. Users are no longer subject to power outages, grid maintenance, or fluctuations in utility prices. This autonomy provides peace of mind and stability.

2. **Environmental Benefits:** Embracing off-grid solar power contributes significantly to reducing the carbon footprint. By generating clean energy from sunlight, users decrease reliance on fossil fuels and lower greenhouse gas emissions, thus actively participating in the fight against climate change.

3. **Remote Accessibility:** Off-grid systems are particularly advantageous in remote or rural areas where connecting to the grid might be expensive or logistically challenging. These systems empower communities that were previously underserved by traditional energy infrastructure.

4. **Long-Term Savings:** Although the initial investment for an off-grid solar system can be substantial, it pays off over time. Users can recoup their costs through energy savings and potentially by selling excess power to others in the area.

Cons:

1. **Higher Initial Costs:** The upfront expenses of purchasing solar panels, batteries, inverters, and other components, along with installation and maintenance costs, can deter individuals from adopting off-grid systems, especially in comparison to grid-tied alternatives.

2. **Complexity of Maintenance:** Off-grid systems require regular monitoring and maintenance. Batteries need consistent attention to ensure optimal performance and longevity. In remote locations, accessing professional maintenance services can be challenging.

3. **Limited Energy Storage:** While advancements in battery technology have improved energy storage capabilities, off-grid systems still rely on the stored energy in batteries. Extended cloudy periods can potentially deplete stored energy, necessitating backup generators or reduced energy consumption.

4. **Sizing Challenges:** Properly sizing an off-grid system requires meticulous consideration of energy consumption patterns, weather conditions, and storage capacity. Undersized systems may lead to energy shortages, while oversized systems result in unnecessary expenses.

Real-Life Off-Grid Solar Power Success Stories

The transition to off-grid solar power systems has engendered remarkable success stories that exemplify the potential and benefits of this alternative energy solution. These anecdotes showcase how off-grid living with solar power can transform lives and communities:

Solar Power Empowers Himalayan Villages

In the remote Himalayan region of Ladakh, India, a network of villages was historically isolated due to challenging terrain and lack of access to modern amenities, including electricity. With the implementation of off-grid solar power systems, these villages now enjoy consistent and clean energy. Solar panels installed on rooftops harness the abundant sunlight, providing lighting, powering appliances, and even facilitating the establishment of small businesses. The project has improved the quality of life and opened avenues for economic growth.

Island Sustainability in American Samoa

On the island of Ta'u in American Samoa, Tesla implemented a groundbreaking off-grid solar and battery solution. Prior to this, the island heavily relied on diesel generators for electricity, which were both environmentally detrimental and expensive. Tesla's installation of solar panels and a battery storage system now fulfills nearly 100% of the island's energy needs. This transformation has significantly reduced the island's carbon emissions and demonstrated the feasibility of large-scale off-grid systems.

Educational Advancement in Rwanda

In Rwanda, the lack of reliable electricity hindered educational progress in rural areas. Off-grid solar power systems, equipped with battery storage, have enabled schools to establish computer labs, provide internet access, and facilitate distance learning. This shift has expanded educational opportunities, bridging the gap between urban and rural students and fostering a brighter future for the nation.

Off-Grid Resilience in Australia

Australia's vast and sparsely populated regions often face challenges in accessing electricity from centralized grids. In response, many Australians have embraced off-grid solar power systems. These systems have provided energy independence and played a crucial role in disaster resilience. During natural disasters that disrupt the main grid, off-grid systems continue to provide power, aiding in communication, medical assistance, and basic necessities.

Off-grid solar power systems represent a dynamic and transformative approach to energy consumption. The differences between grid-tied and off-grid systems underscore the importance of energy autonomy and resilience. While off-grid living with solar power presents both advantages and challenges, real-life success stories illuminate the positive impact this solution can have on individuals, communities, and the environment. As technology continues to advance and costs decrease, the allure of off-grid systems is expected to grow, fostering a more sustainable and decentralized energy landscape.

CHAPTER 2: ADVANTAGES AND BENEFITS OF GOING OFF-GRID

In an era of technological advancement and environmental consciousness, going off-grid has gained substantial traction. Departing from conventional centralized energy systems, the off-grid lifestyle champions self-sufficiency, sustainability, and reduced ecological impact. This chapter delves into the multifaceted advantages and benefits of embracing an off-grid existence, encompassing energy independence, environmental impact mitigation, cost savings, return on investment (ROI), and the liberating off-grid lifestyle.

Energy Independence and Environmental Impact

Energy independence, a hallmark of off-grid living, emerges as a compelling advantage, fostering an individual's or community's self-reliance. Conventional energy sources like fossil fuels are finite, geopolitically sensitive, and contribute to environmental degradation. In contrast, off-grid systems predominantly rely on renewable energy sources such as solar, wind, hydro, and biomass, which are virtually inexhaustible and emit minimal greenhouse gases.

Solar panels, for instance, harness sunlight and convert it into electricity through the photovoltaic effect. Wind turbines capture kinetic energy from the wind to generate power. These technologies can empower individuals and communities to harness nature's abundance, reducing reliance on external energy providers and fostering resilience in the face of energy crises or supply disruptions.

From an environmental perspective, the off-grid approach exerts a significantly reduced carbon footprint. Traditional power generation methods, like coal-fired plants, emit pollutants that degrade air quality and contribute to climate change. By transitioning to off-grid systems, the emission of harmful gases and particulates is curtailed, contributing to cleaner air and a healthier ecosystem. This shift aligns with global efforts to combat climate change outlined in agreements like the Paris Agreement.

Cost Savings and Return on Investment (ROI)

Beyond the philosophical allure of energy independence, off-grid living offers tangible economic benefits. While the initial setup costs of off-grid systems can be substantial, they are often outweighed by the long-term savings they generate. The elimination of monthly energy bills, a hallmark of off-grid living, translates into financial freedom over time. As utility prices continue to fluctuate, the cost predictability of off-grid systems becomes increasingly attractive.

A key financial advantage of the off-grid lifestyle lies in its potential for a strong return on investment (ROI). Solar panels, for instance, have an average lifespan of 25 to 30 years, during which they generate electricity essentially for free. This translates to a considerable ROI, particularly when government incentives and tax credits are factored in. In regions with abundant sunlight, the excess energy generated by off-grid systems can even be sold back to the grid, further enhancing the ROI.

Furthermore, the scalability of off-grid systems permits flexibility in initial investment. Systems can be designed to suit individual needs, from a small cabin in the woods to an entire off-grid community. This adaptability ensures that off-grid living is accessible to a wide range of budgets and circumstances.

Off-Grid Living Lifestyle: Freedom and Sustainability

Beyond the pragmatic advantages, the off-grid lifestyle embodies a profound sense of freedom and sustainability. It enables individuals to break free from the shackles of external energy providers and societal norms, fostering a sense of autonomy. The off-grid mindset encourages resourcefulness, as individuals must devise creative solutions to meet their energy, water, and waste management needs.

Living off-grid also offers a unique opportunity to reconnect with nature. Remote off-grid locations often immerse individuals in pristine environments, allowing them to foster a deeper connection with the natural world. This communion with nature can lead to enhanced mental well-being, reduced stress, and an overall improved quality of life.

Sustainability is a cornerstone of the off-grid lifestyle. The emphasis on renewable energy sources, coupled with conscious consumption patterns, reduces the strain on ecosystems and promotes responsible stewardship of resources. The practice of "living lightly" becomes a tangible reality as individuals strive to minimize their environmental impact.

Additionally, off-grid living encourages self-sufficiency in various aspects beyond energy. Rainwater harvesting, composting toilets, and home gardening are just a few examples of the practices embraced by off-grinders to reduce their reliance on external systems. This holistic approach aligns with the broader global movement towards sustainable living and resilience.

The advantages and benefits of going off-grid extend beyond the surface appeal of energy independence. This lifestyle embodies a transformative shift towards self-sufficiency, environmental responsibility, and financial freedom. The reduction in carbon footprint, the potential for significant ROI, and the nurturing of a sustainable ethos collectively make off-grid living an increasingly attractive proposition. Moreover, the off-grid lifestyle offers an experiential connection with nature and a profound sense of liberty that resonates with those seeking a departure from conventional norms. As societies continue to grapple with the challenges posed by climate change and resource scarcity, the off-grid paradigm stands as a beacon of hope and a tangible pathway towards a more harmonious coexistence with our planet.

Thank you for your purchase!

We extend our sincere gratitude for choosing "The DIY Off Grid Solar Power Bible"

as a part of your reading repertoire.

Scan this QR-CODE to get your **FREE BONUS BOOK** resource carefully curated

to deepen your understanding of essential homesteading practices,

self-sufficiency techniques, and strategies for cultivating a thriving homestead.

CHAPTER 3: ESSENTIAL COMPONENTS OF A SOLAR ENERGY SYSTEM

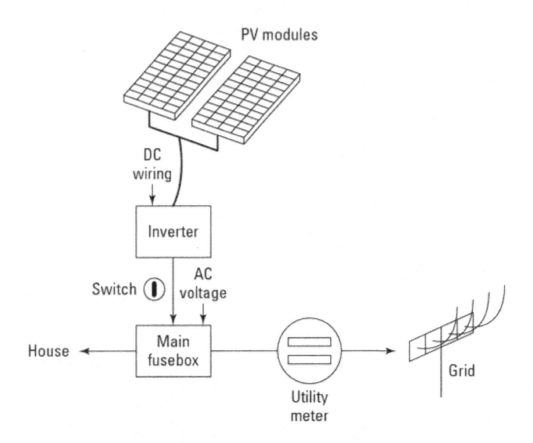

In the pursuit of sustainable and renewable energy sources, solar energy has emerged as a frontrunner, offering an environmentally friendly and efficient solution to the world's energy demands. As societies become increasingly conscious of their carbon footprint, the adoption of solar energy systems has gained remarkable traction. The foundation of any effective solar energy system rests on its essential components, each playing a pivotal role in harnessing, storing, and utilizing solar power. This chapter delves into the intricate details of these components, namely solar panels, deep-cycle batteries, and charge controllers, elucidating their types, efficiencies, and optimal applications.

Solar Panels: Types, Efficiency, and Wattage Calculation

The solar panel, also known as a photovoltaic (PV) module, is at the core of any solar energy system. These panels are responsible for converting sunlight directly into electricity through the photovoltaic effect. Over the years, solar panel technology has evolved, resulting in various types catering to specific needs and environments.

Types of Solar Panels

1. **Monocrystalline Panels:** These panels are constructed from a single crystal structure, making them highly efficient due to their uniform composition. Monocrystalline panels are recognized for their sleek black appearance and are well-suited for installations with limited space.

2. **Polycrystalline Panels:** Made from multiple silicon crystals, polycrystalline panels are easier and cheaper to manufacture. While their efficiency may be slightly lower than monocrystalline panels, they remain popular for residential and commercial installations.

3. **Thin-Film Panels:** This type of panel employs thin layers of photovoltaic materials, making them lightweight and flexible. Thin-film panels, like curved surfaces, are ideal for unconventional installations and are often used in building-integrated photovoltaics (BIPV).

Efficiency of Solar Panels

Solar panel efficiency denotes the percentage of sunlight that a panel can convert into usable electricity. Higher-efficiency panels require less space to generate the same amount of power as lower-efficiency ones. As of the most recent technological advancements, monocrystalline panels tend to boast the highest efficiencies, ranging from 15% to 20%, while polycrystalline panels range from 13% to 16%, and thin-film panels are around 11% to 13%.

Wattage Calculation

The wattage of a solar panel refers to its potential power output under standard test conditions (STC). Calculating the required wattage for a solar energy system involves considering the energy consumption of the devices it will power.

The formula is simple:

Daily Energy Consumption / Peak Sun Hours = Required Panel Wattage.

Peak sun hours refer to the number of hours in a day during which solar irradiance averages 1000 watts per square meter. This value varies based on geographic location and weather conditions. When calculating panel wattage, it is important to account for factors such as shading, panel orientation, and system losses.

Deep-Cycle Batteries: Lead-Acid vs. Lithium-Ion

Efficient energy storage is crucial in any solar energy system to ensure a continuous power supply, even during nighttime or cloudy days. Deep-cycle batteries play a pivotal role in storing excess solar energy for use when the sun isn't shining. Two primary types of deep-cycle batteries are commonly employed in solar setups: lead-acid and lithium-ion.

Lead-Acid Batteries

Lead-acid batteries have been a staple in energy storage for decades. They come in two main varieties: flooded lead-acid and valve-regulated lead-acid (VRLA) batteries. Flooded lead-acid batteries require maintenance, including topping up electrolyte levels, while VRLA batteries are maintenance-free and offer better safety due to their sealed design.

These batteries are cost-effective but have limitations. They are bulkier and heavier compared to lithium-ion batteries, and their depth of discharge (DoD) – the percentage of the battery's capacity that can be used before recharging – is relatively shallow, usually around 50%. This means that a significant portion of the battery's capacity remains unused to prolong its lifespan.

Lithium-Ion Batteries

Lithium-ion batteries have gained immense popularity due to their high energy density, lightweight nature, and deeper depth of discharge, typically around 80% or more. This means more usable capacity for the same battery size. They are also known for their longer lifespan compared to lead-acid batteries.

However, lithium-ion batteries are more expensive upfront, which might deter some users. Additionally, they require a battery management system (BMS) to ensure safe operation, as they can be prone to thermal runaway if not handled properly. While their initial cost is higher, their long-term benefits in terms of performance and lifespan often offset this drawback.

Choosing the Right Battery

Selecting the appropriate battery type depends on the specific requirements of the solar energy system. Lead-acid batteries are cost-effective for smaller installations with relatively low energy storage needs. On the other hand, lithium-ion batteries shine in larger setups where higher energy density, greater efficiency, and longer cycle life are crucial.

Charge Controllers: PWM vs. MPPT and Proper Sizing

Charge controllers, also known as charge regulators, are vital components in solar energy systems that prevent overcharging and damage to batteries by regulating the flow of current from solar panels to the batteries. They come in two main types: Pulse Width Modulation (PWM) and Maximum Power Point Tracking (MPPT) controllers.

PWM Charge Controllers

PWM controllers are an older technology and are generally less expensive compared to MPPT controllers. They work by rapidly connecting and disconnecting the solar panels from the batteries, effectively maintaining a stable battery voltage. However, they are less efficient than MPPT controllers, particularly when dealing with higher voltages and during overcast conditions.

MPPT Charge Controllers

MPPT controllers are the advanced option, utilizing sophisticated algorithms to track the maximum power point of the solar panels in real-time. This allows them to extract the maximum available power from the panels, even under varying weather conditions. As a result, MPPT controllers are more efficient and can provide up to 30% more energy compared to PWM controllers.

Proper Sizing of Charge Controllers

Selecting the right charge controller involves considering the total wattage of the solar panels and the battery bank voltage. The rule of thumb is to choose a charge controller that can handle at least 25% more current than the panels can produce at their maximum output. This ensures that the charge controller can handle potential surges and variations in solar panel performance.

In the realm of solar energy systems, understanding the essential components is paramount to designing and implementing an efficient and effective setup. Solar panels, deep-cycle batteries, and charge controllers form the backbone of these systems, each contributing unique functionalities to ensure a seamless flow of clean and renewable energy. By comprehending the types, efficiencies, and optimal applications of these components, individuals and organizations can harness the power of the sun to pave the way toward a greener and more sustainable future.

CHAPTER 4: CALCULATING YOUR POWER NEEDS AND ENERGY CONSUMPTION

In an increasingly energy-dependent world, understanding and managing our power needs and energy consumption have become vital for both environmental sustainability and financial efficiency. This chapter delves into the intricacies of determining your power requirements, conducting an energy audit, estimating daily energy needs for various applications, and appropriately sizing a solar system to fulfill your energy demands.

Energy Audit: Assessing Your Electricity Usage

Before embarking on any energy-related endeavors, conducting a comprehensive energy audit is imperative. An energy audit involves a systematic analysis of your electricity consumption patterns, helping you identify energy-intensive areas and opportunities for optimization. This assessment empowers you to make informed decisions about energy-saving strategies and investments.

Gathering Data

Begin by collecting historical energy consumption data. Obtain your utility bills for at least the past 12 months, as these records offer insights into your seasonal energy usage fluctuations. Note down the kilowatt-hours (kWh) consumed each month, which will serve as a baseline for your calculations.

Identifying Energy Consumption Patterns

Analyze the data to identify trends and patterns. Categorize your energy consumption into major areas, such as lighting, heating, cooling, appliances, and electronics. This segmentation helps you pinpoint the main contributors to your electricity usage.

Peak Usage Analysis

Pay attention to peak usage periods. These are times when your energy consumption spikes due to the simultaneous operation of high-energy appliances or activities. For instance, if your family tends to use multiple high-energy devices like air conditioners and ovens during the evening, this data will guide your energy management strategies.

Equipment and Appliance Assessment

Inspect your appliances and equipment for energy efficiency. Older models tend to consume more energy than their modern, energy-efficient counterparts. Consider replacing or upgrading appliances that are disproportionately contributing to your energy bills.

Building Envelope Evaluation

Assess the energy efficiency of your home's building envelope. Check for insulation gaps, drafty windows, and other factors that might lead to energy wastage. Improving your home's insulation and sealing can significantly reduce energy demands for heating and cooling.

Recognize the impact of personal behaviors on energy consumption. Simple practices like turning off lights when not needed, unplugging chargers, and using appliances mindfully can collectively make a substantial difference.

For a more comprehensive assessment, consider hiring a professional energy auditor. These experts employ advanced tools such as thermal imaging cameras to detect hidden energy leaks and provide tailored recommendations to enhance your energy efficiency.

Estimating Daily Energy Requirements for Different Applications

Once armed with insights from your energy audit, you can proceed to estimate your daily energy requirements for various applications. This step is crucial for sizing a solar energy system that can adequately meet your demands.

Lighting the Way

As the sun dips below the horizon, illuminating your living space becomes essential. Take a close look at your lighting needs: How many lights do you have, and what are their wattages? When estimating your lighting energy consumption, consider the shift towards energy-efficient LED bulbs. Unlike traditional incandescent bulbs, LED bulbs consume significantly less energy while providing the same brightness. To estimate daily lighting energy usage, the calculation is simple: Multiply the wattage of each bulb by the number of hours they're in operation. Let's say you have four LED bulbs, each with a wattage of 10W, and you use them for an average of 4 hours every day.

LED bulb (10W) x 4 bulbs x 4 hours = 160 Wh (0.16 kWh)

Powering Essential Appliances

Your daily life revolves around appliances, from keeping your food fresh in the refrigerator to working on your laptop. To estimate energy usage, first, determine the wattage of each appliance and how long they're operational. For appliances that cycle on and off, like refrigerators and freezers, you'll need to factor in their usage cycle. The goal is to calculate the average power consumption and then multiply it by the hours of operation. Consider a refrigerator with a wattage of 100W. Since it cycles on and off, let's assume it operates continuously for 24 hours.

Refrigerator (100W) x 24 hours (cycling) = 2400 Wh (2.4 kWh)

Staying Warm

In colder climates, heating systems, such as space heaters, become essential for maintaining comfort. Estimating their energy consumption involves knowing the wattage of your heating device and the number of hours it runs each day.

This calculation provides a clearer picture of how much energy you need to keep yourself warm. Imagine you're using a space heater with a wattage of 1500W for 4 hours each day.

Space heater (1500W) x 4 hours = 6000 Wh (6 kWh)

Charging Electronics

Our modern lives revolve around electronic devices, and ensuring they're charged is a necessity. This category includes smartphones, laptops, tablets, and more. To estimate daily energy consumption, ascertain the wattage of each charger and the duration required for charging. For instance, if you have a laptop charger with a wattage of 65W that's used for 2 hours and a phone charger with a wattage of 10W that's used for 1 hour:

Laptop charger (65W) x 2 hours + Phone charger (10W) x 1 hour = 140 Wh (0.14 kWh)

Appliances with Variable Usage

Certain appliances, like washing machines, aren't used constantly but periodically. Estimating their energy consumption involves calculating the wattage, the hours of use, and factoring in the frequency of usage. This approach provides a more accurate representation of their energy impact. Suppose you use a washing machine with a wattage of 500W for 1 hour each time, and you do laundry three times a week:

Washing machine (500W) x 1 hour (per use) x 3 uses/week = 1500 Wh (1.5 kWh)

By delving into the specifics of your daily energy needs across various applications, you'll be better equipped to calculate your overall energy consumption accurately. This information is the cornerstone for determining the appropriate size and capacity of your solar power system. Keep in mind that your lifestyle may evolve over time, influencing your energy requirements. Regularly revisiting and adjusting these estimations ensures that your solar power system continues to meet your needs effectively.

Sizing Your Solar System to Meet Energy Demand

Armed with a clear understanding of your energy consumption patterns, it's time to determine the size of the solar energy system that can effectively cater to your needs.

1. Solar Irradiance Assessment

The amount of sunlight your location receives, known as solar irradiance, plays a pivotal role in sizing your solar system. Consult solar irradiance maps or use online tools to estimate the average daily sunlight hours your area receives.

2. System Capacity Calculation

Calculate the capacity your solar system needs to generate to meet your energy demands. Divide your total daily energy consumption (in kWh) by the average daily sunlight hours. This will yield the capacity your solar system should have in kilowatts (kW).

3. Accounting for Efficiency

Solar panels and inverters are not 100% efficient, meaning they won't convert all the captured sunlight into usable electricity. Factor in the efficiency of your chosen solar panels and inverters when determining the required system capacity.

4. Future Considerations

While sizing your solar system, contemplate future changes in energy consumption. If you plan to add new appliances or expand your household, accounting for these changes now can save you from the hassle of system upgrades later.

5. Professional Consultation

For precise system sizing, consider consulting with solar energy professionals. They possess the expertise to assess your energy audit data, location, and other specific factors to recommend the ideal solar system capacity for your unique requirements.

Calculating your power needs and energy consumption is a meticulous process that demands attention to detail and understanding your household's energy dynamics. An energy audit forms the foundation, enabling you to identify energy-intensive areas and devise strategies for optimization. Estimating daily energy requirements for different applications is the next step, providing clarity on your consumption patterns. Finally, sizing your solar system bridges the gap between energy demand and supply, ensuring you harness the power of the sun effectively. By undertaking these steps diligently, you contribute to a more sustainable future and potentially reap substantial financial benefits through reduced energy bills and increased energy independence.

BOOK 2

SOLAR POWER SYSTEM DESIGN AND PLANNING

CHAPTER 1: SITE ASSESSMENT AND SOLAR RESOURCE ANALYSIS

Solar power has emerged as a leading contender in a world where environmental concerns and the quest for sustainable energy solutions are at the forefront of technological advancement. The prospect of harnessing the immense power of the sun's rays to generate electricity presents an environmentally friendly alternative and a promising economic avenue for the future. However, the process of installing solar panels goes beyond a mere placement exercise; it entails a meticulous, thorough, and comprehensive analysis of the site and solar resources. This chapter delves deep into the intricacies of site assessment and solar resource analysis, elucidating key aspects such as conducting a site survey, meticulously analyzing local climate data, and judiciously selecting the optimal location for the installation of solar panels.

Conducting a Site Survey: Sunlight Exposure and Obstructions

The cornerstone of any successful solar panel installation project lies in the careful execution of a site survey. This preliminary step involves the physical assessment of the area where the solar panels are slated for installation, with the intention of gaining a comprehensive understanding of the site's unique attributes, potential advantages, and inherent challenges. The focal point of this assessment revolves around two critical considerations: assessing sunlight exposure and identifying potential obstructions that might impede the optimal functioning of the solar panels.

Sunlight Exposure Analysis

At the heart of solar energy generation lies the fundamental principle of sunlight availability. The intensity and duration of sunlight exposure serve as the bedrock for the efficiency of solar panels. During the site survey phase, experts delve into various factors that influence sunlight availability. Geographical location, solar incidence angle, and the prospect of shading caused by structures or natural elements are meticulously evaluated.

The site's geographical location plays a pivotal role in determining the amount of solar energy it receives. Solar panels perform at their peak when they are oriented to receive direct sunlight. Analyzing the solar path, or the trajectory of the sun throughout the year, is crucial for establishing optimal positioning and tilt angles. The solar path also helps in understanding potential variations in energy generation across different seasons.

Obstruction Identification

The impact of shading on solar panels cannot be overstated. Trees, nearby buildings, and other structures cast shadows that can considerably diminish the output of solar panels. Therefore, it is imperative to identify potential sources of shading and assess their potential impact on the solar array's overall performance.

Advancements in technology have led to the utilization of tools like solar pathfinders, which provide a comprehensive visualization of the sun's trajectory throughout the year. These tools offer a precise understanding of areas prone to shading, thereby allowing experts to make informed decisions about panel placement.

Roof Analysis (For Rooftop Installations)

Rooftop installations present unique challenges and opportunities. The roof surface's orientation and tilt significantly influence solar panels' energy production. To maximize sunlight exposure, the azimuth (orientation) of the roof should ideally face south in the northern hemisphere and north in the southern hemisphere. The tilt angle should also align with the geographical latitude for optimal performance.

Roof analysis further extends to assessing the structural integrity of the roof. Solar panels add a considerable load to the roof structure, necessitating a structural assessment to ensure that the roof can withstand the additional weight and various environmental stresses.

Analyzing Local Climate Data for Solar Energy Estimation

The efficacy of solar energy generation is inherently tied to climatic patterns. Solar irradiance, the power per unit area received from the sun, is subject to variations influenced by geographical location and prevailing atmospheric conditions. Analyzing local climate data assumes a pivotal role in providing accurate estimations of potential solar energy generation.

Solar Irradiance Data

Solar irradiance forms the backbone of solar energy generation. It represents the energy received from the sun's rays and varies based on factors such as latitude, altitude, and atmospheric conditions. Historical solar irradiance data specific to the site under consideration serves as a crucial foundation for predicting average energy output over the course of a year.

Solar irradiance data is often derived from comprehensive databases that compile information from various sources, including weather stations, satellites, and ground-based measurements. This data provides insights into seasonal variations in solar energy availability, allowing for more accurate energy production estimates.

Temperature and Heat Effects

Solar panels are sensitive to temperature changes. High temperatures can lead to reduced panel efficiency, as the output of photovoltaic cells tends to decrease as their temperature rises. Analyzing local temperature patterns helps in estimating potential efficiency losses due to elevated temperatures.

This analysis is particularly relevant in regions with extreme temperature fluctuations. Understanding how temperature affects panel efficiency aids in devising cooling mechanisms or selecting panels designed to mitigate the impact of high temperatures.

Weather Patterns

Weather patterns, including cloud cover, precipitation, and atmospheric conditions, can have a significant impact on solar energy generation. Regions with frequent cloud cover or inclement weather may experience fluctuations in energy output. Analyzing long-term weather patterns assists in understanding these variations and enables more accurate estimations of energy generation.

Weather data, often obtained from meteorological stations or satellite observations, provides valuable insights into the frequency and duration of cloudy days, helping stakeholders anticipate potential energy generation fluctuations.

Choosing the Optimal Location for Solar Panel Installation

The selection of the installation location constitutes a critical decision that can significantly influence the overall success and efficiency of a solar panel project. This decision-making process necessitates a holistic evaluation of the results from the site assessment, adherence to local regulations, and alignment with the intended purpose of the solar installation.

Maximizing Sunlight Exposure

Based on the insights gained from the site survey and sunlight exposure analysis, the goal is to position solar panels in locations with the maximum available sunlight exposure and minimal shading. Direct sunlight for a substantial portion of the day is imperative to ensure optimal energy generation.

This step involves the precise placement of solar panels, taking into account azimuth and tilt angles to capture the most sunlight throughout the day. The synergy between the sun's trajectory, potential obstructions, and panel orientation is key to optimizing energy output.

Considering Aesthetics and Regulations

In urban and residential settings, aesthetic considerations often intersect with solar panel installations. Homeowners' associations, local regulations, and zoning ordinances may impose restrictions on the placement and visibility of solar panels. Balancing the imperative of optimal panel placement with compliance with these regulations becomes a delicate task.

The aesthetics of solar panel installations have gained significance, prompting the development of more visually appealing panel designs. Integrating panels seamlessly into the architecture of the building or landscape helps strike a balance between function and aesthetics.

Roof vs. Ground Installation

Solar panels can be installed on rooftops or on the ground, each with advantages and challenges. Rooftop installations are popular due to their space-saving nature, making them ideal for residential and commercial buildings. However, ground installations are better suited for larger arrays and sites with shading issues.

Ground installations offer flexibility in terms of adjusting tilt angles to maximize energy generation. They are particularly suitable for locations with limited rooftop space or suboptimal roof conditions.

Structural Considerations

The chosen installation location's structural integrity is paramount. Solar panels exert additional weight on the structure they are mounted on. Engineering assessments ensure that the selected location can withstand the added load and endure the environmental stresses to which it will be subjected.

Structural considerations encompass load-bearing capacity, wind resistance, and seismic resilience. An in-depth structural analysis helps avoid potential safety hazards and ensures the longevity of the solar installation.

Long-Term Goals

The overarching purpose of the solar installation significantly shapes its design and configuration. Whether the primary objective is to reduce electricity bills, contribute excess energy to the grid, or support off-grid power needs, the intended purpose influences key decisions.

For instance, a residential installation aimed at reducing energy costs might have different design priorities than a utility-scale installation contributing to a regional power grid. Aligning the installation with long-term goals is essential for maximizing the return on investment.

The preliminary stages of a solar panel installation project bear immense significance in the overall success of the endeavor. A comprehensive site assessment and rigorous solar resource analysis lay the foundation for an efficient, effective, and enduring solar energy system. By intricately evaluating factors such as sunlight exposure, potential obstructions, local climate data, and site-specific considerations, stakeholders are empowered to make informed decisions that culminate in optimal energy generation and sustained success. As solar technology continues to evolve, the pivotal role of these preliminary steps remains steadfast, ensuring that our journey toward a sustainable energy future is unequivocally driven by the limitless potential of the sun.

CHAPTER 2: SIZING YOUR SOLAR SYSTEM FOR TINY HOMES

The modern pursuit of sustainability and efficient living has given rise to a novel housing solution - the tiny home. These small, well-designed living spaces offer a unique blend of minimalism and environmental consciousness. With their compact size typically ranging from 100 to 400 square feet, tiny homes have captured the hearts of individuals seeking an alternative way of living that significantly reduces their carbon footprint. Central to the functionality of these eco-friendly havens is the integration of solar power systems, liberating homeowners from dependence on conventional energy grids. This chapter delves into the intricate details of sizing solar power systems specifically tailored for tiny homes.

Understanding Load Profiles for Small Living Spaces

The foundational step in sizing a solar power system for a tiny home involves understanding its load profile comprehensively. This entails analyzing the energy consumption patterns of the dwelling, encompassing all electrical appliances and devices used within it. By examining factors such as power ratings and the duration for which these appliances operate, it becomes possible to accurately quantify the total energy requirement of the tiny home over a given period.

It's important to note that the load profiles of tiny homes differ significantly from those of traditional households. The limited space characteristic of tiny homes naturally results in a reduced number of appliances and lighting fixtures. However, optimizing energy usage is of paramount importance due to the limited energy generation capacity of a compact solar setup. Consequently, the careful selection of energy-efficient appliances is a key consideration.

To effectively evaluate the load profile, the utilization of energy monitoring devices proves to be immensely valuable. These smart tools provide real-time data on energy consumption, enabling homeowners to identify peak and off-peak usage patterns. This data-driven approach not only aids in appropriately sizing the solar system but also encourages conscientious energy consumption practices.

Off-Grid Solar Design Considerations for Tiny Homes

The allure of tiny homes lies not just in their cozy charm but also in their minimalistic footprint on the environment. As you embark on the journey of off-grid solar design for your tiny abode, a unique set of considerations comes into play. Crafting an efficient and effective solar power system for your tiny home requires a thoughtful approach that balances energy needs, available space, and sustainable living principles.

One of the primary considerations in designing an off-grid solar system for a tiny home is the limited available space. Tiny homes, by design, emphasize compact living, which means that finding suitable areas for solar panels is crucial. Rooftop solar panels are a popular choice, maximizing the use of vertical space while minimizing the impact on the home's footprint. Additionally, some tiny home dwellers opt for portable solar panels that can be positioned strategically throughout the day to capture the optimal sunlight.

Energy efficiency takes center stage when dealing with limited space in tiny homes. From LED lighting and energy-efficient appliances to well-insulated windows and doors, every aspect of energy consumption matters. By reducing energy needs through smart design choices, you can not only make the most of your solar system but also create a more sustainable living space overall.

Tiny home living often involves a conscious choice to simplify and downsize. This ethos extends to energy consumption as well. Carefully assessing your daily energy needs is crucial for sizing your solar system appropriately. Understand the power requirements of your appliances, lighting, electronics, and heating systems. By adopting energy-efficient appliances and incorporating passive heating and cooling techniques, you can further optimize your energy usage.

Battery storage is a critical consideration for tiny home solar systems. Since tiny homes have limited roof space for solar panels, the energy generated during peak sun hours needs to be stored efficiently for use during periods of low sunlight. Selecting the right battery capacity ensures that you have a consistent power supply, even on cloudy days or during the night. Lithium-ion batteries are often favored for their high energy density and longer lifespan.

In the context of tiny homes, mobility can also play a role. Some tiny homeowners choose to relocate their homes, which means that the solar system should be designed with portability in mind. Portable solar panels, lightweight batteries, and modular components can make the transition between locations smoother while maintaining your energy independence.

Moreover, embracing off-grid solar power aligns inherently with the philosophy of sustainable living that often accompanies tiny home ownership. Tiny homeowners are already attuned to minimizing their environmental impact, and an off-grid solar system perfectly complements this ethos. By generating clean energy from the sun, you're not only reducing your carbon footprint but also demonstrating the viability of eco-friendly living.

Balancing Power Generation and Energy Consumption

One of the central challenges in sizing a solar system for a tiny home lies in finding the delicate balance between power generation and energy consumption. The goal is to design a system that consistently meets the household's energy demands without overproducing and squandering valuable solar resources.

Achieving this equilibrium requires a thorough analysis of energy usage patterns. A comprehensive understanding of when energy demand peaks and ebbs empowers homeowners to design their solar systems effectively. For instance, if the load profile indicates higher energy consumption during the daytime due to appliances like refrigerators and air conditioning units, the solar system can be tailored to maximize energy production during those hours.

Furthermore, the concept of "net-zero energy" is gaining traction in the realm of sustainable tiny homes. This involves generating as much energy as is consumed over a specific timeframe, often achieved through meticulous system sizing and energy-efficient practices. Energy-efficient appliances, LED lighting, proper insulation, and conscientious consumption all contribute to this ambitious goal.

Technological advancements offer invaluable assistance in this intricate dance between power generation and consumption. Smart energy management systems, equipped with predictive analytics, enable homeowners to anticipate energy usage patterns and optimize their solar systems accordingly. These systems learn from historical data, weather forecasts, and user behavior to make real-time adjustments that ensure a harmonious balance between solar energy production and usage.

Sizing a solar system for a tiny home is a nuanced endeavor that demands a holistic understanding of energy consumption patterns, off-grid system dynamics, and the intricate interplay between power generation and usage. As the world shifts towards more sustainable and mindful living, incorporating solar power in tiny homes is a testament to human ingenuity and environmental stewardship.

CHAPTER 3: DESIGNING SOLAR SYSTEMS FOR CABINS AND REMOTE LOCATIONS

The quest for sustainable and self-sufficient living has ushered in a new era of energy solutions, with solar power taking the forefront. This chapter embarks on an in-depth exploration of the intricate facets involved in designing solar systems tailor-made for the energy needs of cabins situated off-grid and in remote locations far from the reaches of the traditional power grid. This comprehensive analysis delves into the multifaceted challenges posed by seasonal variations in energy production and consumption, the logistical hurdles inherent in transporting and installing equipment in remote locales, and the paramount importance of backup power solutions to sustain energy supply during prolonged periods of low sunlight.

Off-Grid Cabin Energy Needs: Seasonal Variations

The allure of off-grid cabins lies in their symbiotic relationship with nature, offering an escape from the cacophony of urban existence. Yet, the tranquility that characterizes these dwellings comes at the cost of a consistent energy supply. To address this challenge, the design of solar systems for off-grid cabins must encompass a holistic understanding of their energy needs, which are subject to significant fluctuations due to seasonal changes.

During the sun-drenched months of spring and summer, the energy consumption within cabins may be relatively modest. This consumption typically includes basic lighting, refrigeration for food preservation, and intermittent charging of electronic devices. However, the energy equation transforms dramatically as the colder months descend. The introduction of heating systems becomes imperative, inducing a substantial surge in overall energy consumption. Consequently, the foundation of a successful solar system design hinges upon the ability to navigate these seasonal variations with finesse.

Mitigating the effects of seasonal energy variations necessitates a strategic combination of solar panel positioning and battery storage optimization. The positioning of solar panels demands a keen understanding of the sun's path throughout the year. South-facing orientations capture the maximum solar exposure for locations in the northern hemisphere. Additionally, fine-tuning the inclination angle of the panels according to the latitude further maximizes energy absorption.

Yet, the heart of tackling these fluctuations rests within the realm of battery storage. Deep-cycle batteries, purpose-built for enduring prolonged discharge cycles, emerge as the linchpin in ensuring a stable energy supply during lean periods. These batteries act as reservoirs, storing surplus energy generated during periods of peak production for utilization when demand surges, particularly during the energy-intensive winter months. Achieving this equilibrium necessitates meticulous calibration between the number of batteries deployed and the cabin's distinct energy consumption patterns. Advanced energy management systems, equipped with predictive algorithms, can forecast consumption trends and facilitate proactive energy storage, enhancing the solar system's efficiency.

Remote Location Challenges: Transporting and Installing Equipment

The allure of remote locations, with their untouched landscapes and unspoiled vistas, casts an irresistible spell. However, these captivating settings also present an array of challenges when it comes to implementing solar energy systems. The very isolation that renders these locations attractive poses unique hurdles in terms of equipment transportation and installation.

The logistical intricacies of transporting heavy and often bulky solar panels, batteries, and associated hardware to remote sites are daunting. Traditional transportation infrastructure is often absent, necessitating a creative approach. Solutions ranging from employing helicopters for aerial transport to utilizing pack animals or specialized off-road vehicles exemplify the ingenuity required. Each option has its own financial, operational, and environmental considerations. Thus, efficient system design that minimizes the equipment requirement assumes paramount importance.

Installing solar systems in remote locales mandates a flexible approach. The topography of such areas can be rugged, rocky, or prone to extreme weather conditions. The selection of anchoring systems, such as ground mounts or pole mounts, requires meticulous consideration to ensure stability and durability. Moreover, the resilience of the system against the forces of nature, including strong winds, heavy snowfall, and temperature fluctuations, must be at the core of the design. This necessitates the use of robust materials and reinforced designs that can withstand the rigors of such environments.

In some instances, leveraging the expertise of local professionals who understand the region's conditions can be a game-changer. Their insights can guide the selection of appropriate equipment, installation techniques tailored to the site's unique challenges, and maintenance strategies that account for the specific demands of the environment.

Backup Power Solutions for Extended Periods of Low Sunlight

Solar energy, while a renewable resource, is inevitably tethered to the availability of sunlight. Extended periods of cloud cover or reduced sunlight, particularly in high-latitude regions, can potentially lead to energy deficits. This underscores the criticality of backup power solutions in the context of cabins and remote locations where maintaining an uninterrupted energy supply is not just a convenience but a necessity.

One of the prevalent backup solutions involves integrating generators powered by alternative fuels, such as propane or biodiesel. These generators seamlessly activate when the battery storage is depleted, ensuring a continuous and seamless power flow to vital systems. However, this approach introduces its own set of environmental concerns, including emissions and the need for fuel storage.

For a more environmentally conscious approach, incorporating wind turbines alongside solar panels emerges as a compelling solution. Wind patterns often complement low solar energy production periods, ensuring a consistent energy output. Hybrid systems that synergistically harness both solar and wind energies offer the advantage of capitalizing on the strengths of both sources while providing a buffer against the inherent variability of each.

Moreover, the realm of backup power solutions is undergoing a paradigm shift with advancements in energy storage technologies. Hydrogen fuel cells and advanced lithium-ion batteries stand as vanguards of this transformation. These innovative solutions offer prolonged energy storage capacity and reduced maintenance requirements, substantially enhancing the reliability and resilience of off-grid energy systems.

CHAPTER 4: RV AND BOAT SOLAR SYSTEM DESIGN CONSIDERATIONS

In an age of increasing environmental awareness and a growing desire for sustainable living, the integration of solar power systems into recreational vehicles (RVs) and boats has gained significant traction. This trend offers a means of harnessing clean and renewable energy and enhances the overall mobility and convenience of these modes of travel. In this chapter, we delve into the intricate realm of RV and boat solar system design considerations, encompassing topics such as the choice between portable and roof-mounted installations for RV solar panels, the challenges posed by waterproofing and corrosion protection in marine solar systems, and the evolving landscape of mobile solar power solutions for both travel and boating.

RV Solar Panels: Portable vs. Roof-Mounted Installations

The decision to equip an RV with solar panels is a pivotal step towards achieving energy independence while on the road. As such, RV enthusiasts are faced with the fundamental choice between two main installation options: portable and roof-mounted solar panels. Each approach carries distinct advantages and limitations, demanding careful evaluation based on individual preferences and practical considerations.

Portable Solar Panels: Unleashing Flexibility

Portable solar panels have emerged as a versatile solution for RV owners seeking the freedom to optimize energy generation according to prevailing conditions. These panels are typically constructed with lightweight materials and equipped with folding mechanisms, allowing for convenient storage and effortless setup upon reaching a destination. One of the key advantages of portable panels lies in their adjustability; campers can position them at the ideal angle to capture maximum sunlight, a feature particularly valuable in locations with varying sun orientations.

Another noteworthy benefit of portable panels is their potential for easy upgrades. RV enthusiasts can augment their solar setup by acquiring additional panels, a flexible option that accommodates the evolving energy needs of travelers. Additionally, portable panels can be detached and used as standalone charging stations, powering other devices like smartphones, laptops, and portable fridges.

However, the convenience of portability does come with some trade-offs. Portable panels are generally less efficient than their roof-mounted counterparts, owing to their smaller size and lower wattage. This limitation could translate to longer charging times and a potential inability to meet the energy demands of power-hungry appliances. Moreover, the setup and dismantling process might become cumbersome for those constantly on the move, detracting from the overall convenience factor.

Roof-Mounted Solar Panels: Seamlessness and Efficiency

Roof-mounted solar panels offer a solution rooted in integration and efficiency. By affixing solar panels to the RV's roof, travelers can optimize sun exposure without the need for manual adjustments or setups. This seamless integration aligns with the 'set it and forget it's principle, allowing occupants to continuously harness solar energy while concentrating on their journey.

One of the most significant advantages of roof-mounted panels is their capacity for generating a higher output of energy. With larger surface areas available for panel placement, more solar cells can be accommodated, resulting in increased wattage and improved energy production. This makes roof-mounted panels particularly suitable for RV owners with substantial energy requirements, such as those who rely heavily on power-intensive appliances or engage in extended stays off the grid.

Nevertheless, roof-mounted panels are accompanied by a few noteworthy considerations. Firstly, the installation process is relatively complex and often requires professional assistance, potentially leading to increased costs. Secondly, the fixed angle of these panels could lead to suboptimal sun exposure in locations with varying sun positions, affecting overall energy generation. Lastly, the added weight of the panels might impact the RV's aerodynamics and fuel efficiency, a factor to be taken into account, especially during long journeys.

Marine Solar Systems: Waterproofing and Corrosion Protection

As solar energy extends its reach to marine vessels, a distinct set of challenges emerges, primarily centered around the exposure of solar systems to water, humidity, and salt-laden air. Marine solar systems designed for boats and yachts must address these challenges through effective waterproofing and corrosion protection measures to ensure both longevity and safety.

Waterproofing: Defying the Ingress of Moisture

The presence of water is an inherent aspect of marine environments, necessitating robust waterproofing strategies. Solar panels and associated components must be shielded from water infiltration to prevent short circuits, component degradation, and potential hazards to passengers and crew members.

Several approaches are employed to fortify the waterproofing of marine solar systems. Encapsulation through the use of specialized sealants, adhesives, and potting compounds can effectively shield vulnerable connections and components from moisture. Furthermore, the selection of materials with intrinsic water-resistant properties, such as marine-grade junction boxes and UV-stabilized cables, contributes to the system's overall reliability.

Corrosion Protection: Battling the Salt and Humidity

Saltwater environments pose a formidable challenge to the durability of marine solar systems due to their propensity to induce corrosion. Solar panels, wiring, and mounting structures are susceptible to saltwater's corrosive effects and high humidity levels. To counteract these corrosive forces, meticulous material selection and preventive measures are imperative.

Utilizing corrosion-resistant materials is a cornerstone of effective protection. This entails employing aluminum or stainless steel for mounting structures and utilizing connectors and cables with marine-grade corrosion-resistant coatings. Additionally, conformal coatings and galvanic isolators can be applied to vulnerable components to create an additional barrier against the corrosive effects of the marine milieu.

Regular maintenance also plays a pivotal role in the longevity of marine solar systems. Routine inspections, cleaning, and the timely replacement of compromised components are essential practices that safeguard against unforeseen failures. By adopting a proactive approach to maintenance, boat owners can mitigate the potential impact of corrosion and ensure the consistent performance of their solar systems.

Mobile Solar Power Solutions for Traveling and Boating

The evolution of solar power solutions has brought forth a new era of mobile energy independence. With advancements in battery technology and portable solar panels, travelers and boaters can now tap into a wealth of energy sources, allowing them to traverse remote locations without sacrificing essential comforts.

Battery Technology: Enabling Energy Storage

Mobile solar power solutions are intrinsically linked with the advancements in battery technology. Solar panels harvest energy from the sun, which can either be directly utilized to power appliances or stored for later use. Lithium-ion batteries, known for their high energy density and efficiency, have become the backbone of mobile energy storage solutions.

These batteries are compact and lightweight and capable of enduring numerous charge and discharge cycles. This resilience is essential for mobile applications where consistent and reliable energy availability is crucial. Moreover, modern battery management systems ensure the optimal performance of these energy storage units, preventing overcharging, undercharging, and potential thermal hazards.

Portable Solar Panels: Unfolding Freedom

In the realm of mobile solar power, portable panels emerge as the ultimate emblem of freedom and flexibility. These panels are designed with mobility in mind, often featuring folding designs, lightweight materials, and convenient carrying cases. This design ethos enables adventurers to harness solar energy wherever their journey takes them.

The utility of portable solar panels is exemplified in their ability to cater to diverse energy needs. Travelers can deploy these panels to charge smartphones, laptops, cameras, and camping lights. Furthermore, they serve as a lifeline for boaters who require a supplementary power source for navigation equipment, communication devices, and essential appliances during prolonged trips.

Integration and Hybrid Solutions: The Best of Both Worlds

In certain scenarios, the marriage of traditional energy sources with solar power proves to be the most prudent approach. Hybrid solutions, which combine solar panels with conventional generators or engine alternators, offer a balanced compromise between energy efficiency and availability.

In boats, for instance, hybrid systems can harness solar energy during daytime hours while relying on conventional sources during the night. Similarly, RV owners can benefit from integrated systems that draw power from both solar panels and the vehicle's engine. These hybrid setups provide a safety net against unforeseen energy demands while capitalizing on the clean energy harnessed from the sun.

The integration of solar power systems into RVs and boats is emblematic of a harmonious coexistence between human mobility and environmental responsibility. The decisions pertaining to RV solar panel installations, the intricacies of marine solar systems, and the evolution of mobile solar power solutions all underscore the intricate dance between innovation, practicality, and sustainability.

Whether opting for the adaptability of portable panels or the efficiency of roof-mounted installations, RV enthusiasts are poised to redefine their travel experiences through the liberation of self-sustaining energy. Marine solar systems, fortified against the elements through waterproofing and corrosion protection measures, offer a testament to human ingenuity's conquest over nature's challenges. Finally, the burgeoning landscape of mobile solar power solutions paints a vivid picture of energy freedom, enabling adventurers and boaters to explore the farthest reaches of our planet while remaining tethered to the clean embrace of the sun.

BOOK 3

SELECTING THE RIGHT SOLAR EQUIPMENT

CHAPTER 1: CHOOSING THE BEST SOLAR PANELS FOR YOUR SETUP

As the world continues to shift towards renewable energy sources, solar power stands at the forefront as a promising and sustainable solution to our energy needs. Harnessing energy from the sun not only reduces our carbon footprint but also provides an opportunity for individuals and businesses to generate their own electricity. Solar panels are central to any solar energy system, which converts sunlight into usable electrical energy. However, with the market flooded with various options, choosing the best solar panels for your setup can be daunting. In this chapter, we will delve into the intricacies of solar panels, comparing different types, evaluating efficiency and performance, and understanding factors that influence their lifespan and maintenance.

Monocrystalline, Polycrystalline, and Thin-Film Panels Comparison

When it comes to solar panels, there are three primary types that dominate the market: monocrystalline, polycrystalline, and thin-film panels. Each type has its own set of advantages and disadvantages, catering to different preferences and requirements.

Monocrystalline Panels

Monocrystalline solar panels are often hailed as the most efficient and space-saving option. They are made from a single continuous crystal structure, allowing them to convert sunlight into electricity at a higher efficiency rate than other types. These panels have a uniform black color and a sleek design, making them visually appealing for residential installations. Additionally, monocrystalline panels perform exceptionally well in low-light conditions, making them suitable for regions with varying weather patterns. However, the advanced technology and higher efficiency come at a slightly higher cost, which should be factored into your budget.

Polycrystalline Panels

Polycrystalline solar panels are a popular choice due to their cost-effectiveness. They are made from multiple silicon fragments, which are melted together to form the panel. While their efficiency is slightly lower than monocrystalline panels, advancements in technology have narrowed this gap. Polycrystalline panels have a bluish hue and a less uniform appearance compared to monocrystalline panels. They also require more space to produce the same amount of energy as monocrystalline panels, which might be a consideration for those with limited installation space.

Thin-Film Panels

Thin-film solar panels are known for their flexibility and versatility. They can be integrated into various surfaces, such as roofs and walls, and are often used in building-integrated photovoltaics. Thin-film panels are made by depositing a thin layer of photovoltaic material onto a substrate like glass or metal. While their efficiency is generally lower than crystalline panels, they perform better in high-temperature environments. Thin-film panels are suitable for unconventional installations or situations where aesthetics play a crucial role. However, they typically require a larger area to generate the same amount of electricity as crystalline panels.

Evaluating Solar Panel Efficiency and Performance Ratings

The efficiency of a solar panel refers to its ability to convert sunlight into electricity. It is crucial to consider when choosing the right solar panels for your setup. Solar panel efficiency is usually expressed as a percentage and measures how much sunlight that strikes the panel's surface is converted into usable energy. Higher-efficiency panels generally produce more electricity in the same amount of sunlight, making them a desirable choice for those looking to maximize their energy output.

Performance ratings also play a pivotal role in determining the suitability of solar panels for your specific needs. Two key metrics to look for are the temperature coefficient and the power tolerance.

Temperature Coefficient

Solar panels are most efficient at lower temperatures. As the temperature rises, their efficiency tends to decrease. The temperature coefficient indicates how much the panel's efficiency will drop as the temperature increases by one degree Celsius. A lower temperature coefficient is indicative of better performance in hot climates. When comparing different panels, it's advisable to opt for panels with lower temperature coefficients, especially if you live in a region with high average temperatures.

Power Tolerance

Power tolerance represents the acceptable range within which the panel's actual output may vary from its rated output. For example, a panel with a +5% power tolerance can produce up to 5% more electricity than its rated output, while a -5% tolerance indicates it might produce up to 5% less. Choosing panels with a smaller power tolerance ensures more accurate energy production estimates and reduces the risk of falling short on your energy needs.

Factors Affecting Solar Panel Lifespan and Maintenance

Investing in solar panels is a long-term commitment, and understanding the factors that influence their lifespan and the maintenance they require is essential to ensure optimal performance over the years.

Quality of Materials

The quality of the materials used in manufacturing solar panels directly affects their longevity. High-quality panels are made using durable and corrosion-resistant materials that can withstand the elements. Choosing panels from reputable manufacturers known for their commitment to quality and reliability is crucial.

Weather Conditions

Solar panels are exposed to various weather conditions throughout their lifespan, including rain, snow, hail, and extreme temperatures. While most panels are designed to withstand these conditions, it's important to consider the specific climate of your region. Panels with higher durability ratings and appropriate weatherproofing are more likely to have a longer lifespan.

Regular Maintenance

Solar panels are generally low-maintenance, but some degree of upkeep is necessary to ensure their efficiency and longevity. Regular cleaning of dirt, debris, and bird droppings is recommended to prevent shading that can affect energy production. Cleaning intervals may vary based on your location and the prevalence of environmental factors that could dirty the panels.

Inverter Maintenance

Inverters are a crucial component of a solar energy system as they convert the direct current (DC) generated by the panels into alternating current (AC) used for household electricity. These inverters typically have a shorter lifespan than the panels themselves. Regular checks and potential replacements are necessary to ensure the overall efficiency of your solar setup.

Shading and Placement

Proper placement of solar panels plays a significant role in their performance and longevity. Even partial shading on a single panel can significantly reduce the energy output of an entire array. Therefore, it's important to install panels in locations where they receive maximum sunlight throughout the day, avoiding potential shading from trees, neighboring buildings, or other obstacles.

Warranty and Support

When selecting solar panels, carefully review the manufacturer's warranty and customer support services. A longer warranty period often reflects the manufacturer's confidence in the panel's durability and performance. Be sure to understand the terms and conditions of the warranty, including coverage for defects, degradation, and other potential issues.

Choosing the best solar panels for your setup involves careful consideration of various factors, from the type of panels to their efficiency, performance ratings, lifespan, and maintenance requirements. Monocrystalline, polycrystalline, and thin-film panels each offer unique advantages, allowing you to tailor your choice to your specific preferences and circumstances. Evaluating efficiency and understanding performance metrics like temperature coefficients and power tolerances ensures you make an informed decision that aligns with your energy needs. Furthermore, recognizing the factors that influence solar panel lifespan and the importance of regular maintenance will contribute to the long-term success of your solar energy investment. As the renewable energy landscape continues to evolve, selecting the right solar panels lays the foundation for a sustainable and efficient energy generation system.

CHAPTER 2: EVALUATING BATTERY OPTIONS FOR ENERGY STORAGE

In the dynamic realm of energy storage, the selection of battery technologies plays a pivotal role in determining the efficiency, reliability, and overall cost-effectiveness of energy storage systems. As the demand for efficient energy storage solutions continues to surge, the significance of comprehending and evaluating various battery options becomes increasingly paramount. This chapter embarks on an in-depth exploration of the critical considerations in evaluating battery energy storage options. Specifically, we delve into the distinctive characteristics, benefits, and considerations associated with two major categories of batteries: Lead-Acid Batteries and Lithium-Ion Batteries. Moreover, we delve into the intricate domain of battery capacity sizing and system voltage configurations, elucidating the intricate interplay among these factors in designing energy storage systems that deliver optimal performance and efficacy.

Lead-Acid Batteries: Deep-Cycle vs. Flooded vs. AGM

With their long-standing presence in the energy storage landscape, lead-acid batteries have earned their reputation as reliable workhorses. They come in diverse types, each tailored to meet specific operational requirements while presenting distinct advantages and trade-offs.

Deep-Cycle Lead-Acid Batteries

Deep-cycle batteries emerge as a quintessential solution for applications necessitating prolonged and consistent energy delivery. These batteries are meticulously engineered to endure repetitive discharges to a greater depth compared to their counterparts. This design characteristic positions them as ideal choices for scenarios where a sustained and steady power output is essential, such as in renewable energy systems. The subtypes of deep-cycle batteries, encompassing Flooded, AGM, and Gel batteries, further provide specialized solutions catering to unique demands.

Flooded Lead-Acid Batteries

Among the lead-acid battery options, flooded lead-acid batteries, often referred to as wet-cell batteries, stand as traditional yet robust contenders. They comprise lead plates submerged in a liquid electrolyte solution, typically composed of sulfuric acid and water. Their cost-effectiveness and dependable performance make them a favored choice. However, these batteries entail regular maintenance obligations. Due to the evaporation that occurs during operation, periodic checks and electrolyte refills are essential to uphold their efficiency.

AGM (Absorbent Glass Mat) Lead-Acid Batteries

Advancing the lead-acid technology, AGM batteries introduce a higher level of sophistication. A fiberglass mat infused with electrolyte is used to immobilize the electrolyte, eliminating the need for a liquid electrolyte flow. The merits of AGM batteries encompass a maintenance-free design, heightened resistance against vibrations, and exceptional performance even under cold conditions. These attributes render AGM batteries particularly well-suited for remote applications and off-grid setups.

Considerations for Lead-Acid Batteries

When embarking on the evaluation of lead-acid battery options, a gamut of factors necessitates careful scrutiny. Parameters such as cycle life, maintenance prerequisites, cost implications, and environmental ramifications all come into play. Flooded batteries underscore cost-effectiveness and robust performance, albeit demanding regular maintenance. AGM batteries counterbalance these virtues with a maintenance-free operation but do entail a slightly elevated upfront cost. The selection among these options hinges on the energy storage system's specific exigencies and the available budgetary provisions.

Lithium-Ion Batteries: Benefits, Types, and Considerations

Innovation takes center stage with the introduction of lithium-ion batteries, ushering in a transformative era in the energy storage landscape. Renowned for their remarkable energy density, lightweight design, and extended cycle life, lithium-ion batteries have emerged as frontrunners in a multitude of applications, spanning portable electronics to electric vehicles and grid-connected energy storage systems.

Benefits of Lithium-Ion Batteries

The towering advantage of lithium-ion batteries resides in their unparalleled energy density. This distinctive characteristic empowers them to encapsulate a substantial quantum of energy within a compact and lightweight form factor. Consequently, this makes them an impeccable choice for installations constrained by space considerations. Moreover, lithium-ion batteries exhibit an extended cycle life, surpassing that of traditional lead-acid batteries. This longevity translates into the capability to withstand a higher count of charge and discharge cycles before encountering a significant capacity decline.

Types of Lithium-Ion Batteries

The kingdom of lithium-ion batteries comprises an array of types meticulously tailored to cater to specific applications and performance requisites:

Lithium Cobalt Oxide (LiCoO2)

LiCoO2 batteries shine with their high energy density, finding prevalent usage in consumer electronics like smartphones and laptops. However, they exhibit reduced stability at elevated temperatures and have a shorter overall lifespan when juxtaposed with other lithium-ion battery variants.

Lithium Iron Phosphate (LiFePO4)

LiFePO4 batteries ascend as exemplars of safety, stability, and durability. Despite a slightly lower energy density than LiCoO2 batteries, they flaunt the ability to endure a broader spectrum of temperatures without compromising performance. This makes them the go-to choice for applications wherein safety and robustness, such as renewable energy storage and electric vehicles, take precedence.

Lithium Manganese Oxide (LiMn2O4)

LiMn2O4 batteries strike a harmonious equilibrium between energy density and safety considerations. They find their niche in power tools and medical devices, credited to their stable chemistry and commendable performance across diverse temperature scenarios.

Lithium Nickel Cobalt Aluminum Oxide (LiNiCoAlO2)

Labeled as NCA batteries, this variant boasts a heightened energy density and a broader operating temperature range when compared to LiCoO2 batteries. The automotive domain, specifically electric vehicles, embraces NCA batteries, where the delicate balance between energy density and thermal stability aligns with automotive prerequisites.

Considerations for Lithium-Ion Batteries

While lithium-ion batteries' advantages are conspicuous, many considerations warrant judicious evaluation. Effective thermal management emerges as a critical consideration to forestall overheating scenarios. Improper handling could potentially trigger thermal runaway, underscoring the significance of prudent operational practices. Moreover, the upfront cost associated with lithium-ion batteries tends to be higher in comparison to traditional lead-acid alternatives. Nevertheless, the elongated lifespan and unparalleled performance often outweigh these initial costs, particularly in applications mandating high energy density and frequent charge-discharge cycles.

Battery Capacity Sizing and System Voltage Configurations

The meticulous determination of battery capacity and the adept configuration of system voltage constitute foundational steps in architecting an energy storage solution that aligns with distinct performance objectives.

Battery Capacity Sizing

Battery capacity sizing encompasses the intricate process of selecting an appropriate quantum of energy storage. This selection aims to ensure that the energy storage system is proficient in delivering ample power during peak demand while upholding operational stability during phases of reduced energy production. The factors influencing capacity sizing span the consumption pattern of energy, projected load demand, and the desired depth of discharge (DOD) that the system will endure.

System Voltage Configurations

The configuration of system voltage, a facet often sculpted by the alignment of batteries in series or parallel connections, bears substantial implications for the overall efficacy and efficiency of the energy storage system. Series connections usher in a voltage upsurge while retaining constant capacity, rendering them apt for applications that mandate higher voltage output. In contrast, parallel connections retain system voltage while augmenting overall capacity, an advantageous attribute for scenarios necessitating extended operation during periods characterized by low energy production.

In the ever-evolving vista of energy storage, the discerning selection of battery options transcends the realm of mere choice. It metamorphoses into a strategic decision that permeates efficiency, sustainability, and operational viability. The spectrum of lead-acid batteries, extending from conventional flooded batteries to sophisticated AGM designs, amplifies the significance of reliability and adaptability. Conversely, lithium-ion batteries redefine the boundaries of energy storage possibilities with their revolutionary attributes of high energy density, elongated cycle life, and an array of chemistry variations. Both these categories, lead-acid and lithium-ion, orchestrate their unique orchestras of advantages and intricacies, necessitating meticulous analysis to ascertain alignment with system prerequisites.

Furthermore, the precision of battery capacity sizing and the astuteness of system voltage configurations stand as cardinal prerequisites in the creation of energy storage solutions. These components emerge as the bedrock upon which the efficacy and performance of the entire system are sculpted. Their symbiotic interplay unearths a symphony wherein every note resonates with optimal energy utilization and seamless operation goals.

As the energy storage domain hurtles forward, propelled by innovation and evolving needs, the cognizant decisions pertaining to battery options, capacity sizing, and voltage configurations etch the trajectory toward a sustainable energy future. This trajectory, punctuated by precision, foresight, and technological acumen, strives to harmonize the symbiotic relationship between human needs and the resources of the planet.

CHAPTER 3: INVERTERS AND CHARGE CONTROLLERS: TYPES AND SELECTION

The world of renewable energy has witnessed remarkable advancements over the past few decades. Solar power has emerged as a leading contender in the race towards sustainable energy solutions as we strive to reduce our reliance on conventional fossil fuels. Solar energy systems, particularly in off-grid scenarios, rely heavily on inverters and charge controllers to ensure seamless energy conversion and efficient utilization. In this chapter, we delve into the intricate details of these vital components, exploring their types, benefits, and the art of selecting the right ones for specific applications.

Off-Grid Inverters: Pure Sine Wave vs. Modified Sine Wave

When it comes to off-grid solar setups, inverters play a pivotal role in transforming the direct current (DC) generated by solar panels into alternating current (AC) – the type of electricity used in most of our electrical appliances. In this context, two primary types of inverters have gained prominence: pure sine wave and modified sine wave inverters.

Pure Sine Wave Inverters

Pure sine wave inverters stand as a testament to engineering precision. They produce a waveform that mirrors the utility grid's AC waveform, ensuring a seamless transition between solar-generated power and conventional grid power. This characteristic makes pure sine wave inverters ideal for sensitive electronic devices, such as computers, medical equipment, and high-end audio systems. The smooth, undistorted waveform guarantees that these devices operate efficiently and reliably without the risk of overheating or malfunction.

One of the standout advantages of pure sine wave inverters is their ability to prevent the infamous "inverter noise" often associated with modified sine wave inverters. This noise can manifest as an audible hum or audio and video equipment interference. The clean power output of pure sine wave inverters eliminates this issue, making them indispensable in applications where quality power is non-negotiable.

Modified Sine Wave Inverters

While pure sine wave inverters offer impeccable waveform replication, modified sine wave inverters take a more pragmatic approach. They approximate the AC waveform through a series of steps, resulting in a stepped waveform rather than a smooth curve. This approach allows modified sine wave inverters to be more cost-effective and efficient in certain applications.

Modified sine wave inverters find their niche in scenarios where the precision of the waveform isn't paramount. They work well with simpler electronics, such as basic household appliances like fans, incandescent light bulbs, and power tools. However, their use with sensitive electronics is not recommended, as the waveform distortion can lead to reduced efficiency and potential long-term damage.

MPPT Charge Controllers: Advantages and Optimal Use Cases

As solar panels generate electricity from sunlight, the voltage and current they produce can vary significantly based on factors like sunlight intensity and temperature. Charge controllers are the unsung heroes that regulate the power flow between solar panels and batteries, ensuring that the batteries receive the right charge and are not overcharged or discharged. Among charge controllers, Maximum Power Point Tracking (MPPT) controllers stand out for their efficiency and adaptability.

Advantages

MPPT charge controllers operate on a sophisticated algorithm that enables them to find the optimal operating point of the solar panels – the maximum power point (MPP). Unlike traditional Pulse Width Modulation (PWM) controllers, which simply turn the power flow on and off, MPPT controllers actively adjust the current and voltage to match the battery's needs while maximizing energy conversion.

One of the primary advantages of MPPT controllers is their ability to handle higher-voltage solar panels and convert that excess voltage into usable current. This means that in situations where solar panels generate more voltage than the battery's current requirements, MPPT controllers can harvest more power compared to PWM controllers. This feature is particularly beneficial in low-light conditions or when panels are wired in series to achieve higher voltages.

Optimal Use Cases

MPPT charge controllers shine in scenarios where space is limited and energy conversion efficiency becomes crucial. Remote installations, such as cabins or RVs, can greatly benefit from MPPT controllers, as they ensure that every ray of sunlight is harnessed to its full potential. Additionally, larger solar installations with multiple panels strung together will benefit from the ability of MPPT controllers to work with varying panel voltages efficiently.

It's important to note that while MPPT charge controllers offer impressive advantages, their higher costs must be considered in the overall system budget. The benefits they provide might outweigh the upfront cost in the long run, but the economic feasibility should be evaluated on a case-by-case basis.

Sizing Inverters and Charge Controllers for Efficient Energy Conversion

Selecting the right size of inverters and charge controllers is a delicate dance that balances power requirements, system capacity, and budget considerations. Oversizing or undersizing these components can lead to inefficiencies, reduced system lifespan, or unnecessary costs. Careful sizing requires a deep understanding of the power demands and available solar resources.

Inverter Sizing

When sizing an inverter, it's essential to match its capacity to the maximum power demand of the appliances it will be powering simultaneously. However, a common mistake is to base the inverter size solely on peak loads, disregarding the fact that many appliances don't always operate at full power. A balanced approach involves analyzing the usage patterns and selecting an inverter that can comfortably handle regular loads while accommodating occasional peaks.

Moreover, considering the inverter's efficiency is critical. Inverters operate at their peak efficiency when loaded to around 70-80% of their capacity. Oversizing the inverter significantly beyond your needs can lead to lower efficiency during typical usage, negating potential benefits.

Charge Controller Sizing

Charge controller sizing revolves around the solar panel array's current and voltage output and the battery bank's capacity. For MPPT controllers, selecting a model that can handle the total panel voltage at the lowest expected temperature ensures that the system doesn't trip into a lower charging state during cold conditions.

Additionally, the charge controller's current rating should align with the total current output of the solar panels. This ensures that the controller can handle the maximum current generated during peak sunlight hours without becoming a bottleneck. Oversizing the charge controller also provides room for future panel additions without requiring a controller upgrade.

CHAPTER 4: OTHER COMPONENTS: CABLES, MOUNTING SYSTEMS, AND MONITORING DEVICES

In the dynamic landscape of renewable energy, solar power stands out as a pivotal player in the race toward sustainable solutions for our growing energy needs. Harnessing energy from the sun's rays requires an intricate interplay of various components, each playing a crucial role in ensuring maximum efficiency, safety, and durability of solar power systems. While solar panels take center stage as the visible face of these systems, there are several behind-the-scenes components that deserve equal attention. This chapter delves into the often-overlooked yet indispensable elements: solar cables and connectors, mounting systems, and monitoring devices.

Solar Cables and Connectors: Proper Sizing and Protection

In the intricate web of an off-grid solar power system, solar cables and connectors are the lifelines that ensure the seamless flow of energy from your panels to your appliances. These components might not always steal the spotlight, but their proper sizing, quality, and protection are vital for maintaining the efficiency, safety, and longevity of your solar setup.

Sizing Solar Cables

Choosing the right size of solar cables is crucial for minimizing energy losses and ensuring the efficient transmission of power. The primary factor to consider is the electrical resistance of the cables. When current flows through a cable, it encounters resistance, leading to the generation of heat and energy loss. Larger cables have lower resistance, allowing more energy to flow without significant losses. The goal is to keep the voltage drop, caused by resistance, to a minimum.

To determine the appropriate cable size, you need to consider the distance the energy needs to travel (cable length), the current (amperage) your system requires, and the voltage of your system. You can use the "voltage drop formula" to calculate the voltage drop for a given cable size and length. The National Electrical Code (NEC) provides guidelines for allowable voltage drop percentages, typically ranging from 1% to 3%, depending on the application.

Choosing Connectors

Connectors are the pivotal junctions that link different components of your solar system. Quality and compatibility are paramount. Poor-quality connectors can introduce resistance, overheating, and even system failures. Look for connectors that are designed for outdoor use, as they need to withstand varying weather conditions.

Matching the connectors to your cables and other components is vital. Connectors come in various sizes and types, such as MC4 connectors, which are prevalent in solar systems. Ensuring that connectors are securely fastened, weatherproof, and able to handle the expected current is essential. Quality connectors provide a seamless transition, allowing energy to flow smoothly through your system.

Cable Length and Voltage Drop

Cable length is a critical factor affecting energy losses. Longer cables introduce more resistance, leading to higher voltage drops and energy wastage. It's advisable to keep cable lengths as short as possible to minimize energy losses. When planning your system layout, consider the distance between solar panels, batteries, and other components.

Protection Against the Elements

Solar cables and connectors are exposed to the elements day in and day out. UV radiation, temperature fluctuations, moisture, and physical wear can degrade their performance over time. UV-resistant cable insulation and corrosion-resistant connectors are essential for maintaining both the electrical and structural integrity of your system.

Using conduit or cable trays can provide additional protection against physical damage. These solutions shield cables from accidental impacts, rodents, and environmental factors, ensuring the longevity of your cables and connectors.

When dealing with solar cables and connectors, safety is paramount. Proper grounding, insulation, and protection against electrical hazards are crucial to prevent accidents and system malfunctions. Following electrical codes and guidelines, such as those outlined by the NEC, ensures that your system meets safety standards and operates reliably.

Mounting Solar Panels: Roof, Ground, Pole, and Tracking Systems

While solar panels are designed to absorb sunlight and convert it into energy, their positioning and support structure play a crucial role in their effectiveness. Mounting systems provide the necessary foundation for solar panels, and choosing the right type depends on various factors, including location, available space, and energy goals.

Roof-Mounted Solar Panels

Roof-mounted solar panels are a popular choice for many homeowners due to their efficient use of space. When installing panels on your roof, several factors come into play. The angle of your roof and its orientation towards the sun determine the optimal placement. In northern latitudes, south-facing roofs are ideal, while in southern latitudes, north-facing roofs can be more effective. Roof angle also influences how well panels capture sunlight. However, keep in mind that roof installations might require additional precautions for sealing and waterproofing to prevent leaks.

Ground-Mounted Solar Arrays

Ground-mounted solar panels offer more flexibility in terms of placement and orientation. These arrays are often used when rooftops are unsuitable due to shading, orientation, or structural concerns. When installing ground-mounted panels, consider the tilt angle for maximizing solar exposure. You can adjust the tilt according to your latitude to capture the optimal amount of sunlight throughout the year. Proper anchoring and foundation work are essential for stability and longevity, particularly in areas prone to harsh weather conditions.

Pole-Mounted Solar Arrays

Pole-mounted solar arrays provide elevation advantages, allowing panels to avoid obstructions such as tall buildings or trees. This approach is particularly useful for remote locations or places where shading is a concern. The height and angle of the pole are critical for achieving optimal sun exposure. Ensure that the pole is sturdy and well-anchored to withstand wind and weather.

Tracking Systems

Tracking systems take solar panel efficiency to the next level by allowing panels to follow the sun's path throughout the day. Single-axis trackers follow the sun's east-to-west trajectory, while dual-axis trackers also adjust for seasonal changes in the sun's angle. Tracking systems can significantly increase energy production but come with higher costs and maintenance requirements. These systems are ideal for large installations where maximizing energy output is a priority.

Balancing Efficiency and Practicality

When deciding on the ideal mounting system, consider a balance between efficiency, practicality, and budget. Rooftop installations are space-efficient but may require structural assessments and maintenance considerations. Ground-mounted arrays offer more flexibility but require suitable land and foundation work. Pole-mounted systems provide elevation advantages, and tracking systems offer the highest efficiency but come with increased complexity.

Regardless of the mounting method you choose, safety is paramount during installation. Securely fastening panels, using proper seals and brackets, and following local building codes ensure a reliable and safe installation. Regular inspections and maintenance checks are vital to keep your system in optimal condition over the long term.

Monitoring and Control Systems for Real-Time Performance Tracking

In the realm of solar power, knowledge is power, and this knowledge is harnessed through advanced monitoring and control systems. These systems provide real-time insights into the performance of the solar power setup, enabling timely interventions and optimizations.

- **Performance Monitoring**: Monitoring systems continuously gather data on various aspects of the solar power system. This includes parameters such as energy production, panel efficiency, and overall system health. Anomalies can be quickly detected and addressed by comparing the actual performance with the expected performance.

- **Fault Detection**: Early detection of faults is essential to prevent downtime and energy loss. Monitoring systems can identify issues such as malfunctioning panels, damaged cables, or connectivity problems. This information allows maintenance teams to proactively address issues before they escalate.

- **Remote Management**: Many modern solar power installations are equipped with remote monitoring capabilities. This means that system data can be accessed and analyzed from a centralized location. This is particularly valuable for large-scale installations or systems located in remote areas.

- **Data Analytics and Trending**: The data collected by monitoring systems can be harnessed for deeper insights. Data analytics can reveal patterns and trends in energy production, consumption, and system behavior. This information is invaluable for making informed decisions about system upgrades, expansions, or optimizations.

- **Performance Optimization**: Monitoring systems provide insights into current performance and contribute to system optimization. By analyzing energy production and consumption data, operators can fine-tune the system for better efficiency and energy utilization.

In the grand tapestry of solar power systems, these often-overlooked components weave a tale of functionality, reliability, and efficiency. Each component plays a pivotal role, from the unpretentious solar cables that channel energy to the intricate dance of mounting systems under the sun's gaze and the vigilant watch of monitoring systems. As the world pivots towards sustainable energy solutions, these behind-the-scenes heroes continue to evolve, ensuring that solar power realizes its full potential on the global stage.

BOOK 4

INSTALLATION AND SETUP OF SOLAR SYSTEMS

CHAPTER 1: SAFETY PRECAUTIONS AND INSTALLATION GUIDELINES

The world of energy is evolving, and the adoption of solar power systems is growing at an unprecedented rate. Harnessing the power of the sun offers a clean and sustainable energy solution, but it's crucial to approach the installation process with diligence and safety in mind. This chapter delves into the paramount importance of safety precautions and installation guidelines when working with solar components. It encompasses various considerations, from electrical safety measures during installation to adhering to local building codes and permit requirements. Furthermore, it expounds on the imperative of roof and ground installation safety practices, highlighting the intricate details that contribute to a secure and efficient solar power system setup.

Electrical Safety Measures: Working with Solar Components

The cornerstone of any solar power installation project lies in the proper handling of electrical components. As solar power involves the conversion of sunlight into electricity, the equipment used inherently carries electrical risks. Ensuring the safety of both installers and future users of the system necessitates stringent adherence to a set of electrical safety measures.

Grounding and Wiring Integrity

A network of electrical components is at the heart of every solar power system. These components, ranging from solar panels to inverters and batteries, are not only the lifeblood of the system but also potential sources of electrical hazards. The bedrock of electrical safety begins with proper grounding and wiring integrity. This practice safeguards against the risk of electric shock, equipment damage due to lightning strikes, and power surges.

A meticulous approach to grounding entails connecting all solar components to the grounding electrode system of the building. This establishes a conductive path for electrical faults to dissipate harmlessly into the ground. Moreover, local regulations and industry standards mandate specific grounding methods and conductors to be employed. Adhering to these guidelines is non-negotiable in ensuring the longevity and safety of the installation.

Wiring within the building and connecting the solar components is equally critical. Insulating wires properly and protecting them from physical damage minimizes the chances of short circuits and electrical fires. Regular inspections of the wiring system are essential to detect any wear and tear, ensuring that the integrity of the system is maintained over its operational life.

DC Voltage Awareness

Understanding the behavior of direct current (DC) voltage is pivotal when working with solar power systems. Unlike the alternating current (AC) voltage typically used in homes, DC voltage is inherent to solar installations. Installers must exercise caution and be acutely aware of the presence of DC voltage throughout the system.

Proper labeling and clear identification of DC lines and components are imperative to mitigate the risk associated with DC voltage. These labels serve as visual reminders, ensuring that installers and maintenance personnel can easily distinguish between AC and DC circuits. This distinction is crucial for safe handling, maintenance, and troubleshooting.

Disconnect Procedures

The cyclical nature of solar power systems demands maintenance and occasional repairs. However, engaging in these activities without adhering to proper disconnect procedures can lead to disastrous consequences. Each solar component should be equipped with accessible disconnect switches that effectively isolate the system from the grid and prevent electrical backfeed.

In practice, this means that installers must follow a meticulous protocol for system shutdown and isolation before any maintenance work. This protects them from potential electrical shocks and ensures that the system remains safely disconnected from the grid, avoiding any power surges that could damage the equipment. This procedural diligence is the cornerstone of responsible solar installation and maintenance.

Roof and Ground Installation Safety Practices

As solar power systems become more prevalent, their installations traverse a multitude of terrains and environments. Two key settings – rooftop and ground installations – demand specific safety practices to ensure installers' well-being and the systems' integrity.

Rooftop Installations

The idea of harnessing sunlight from rooftops embodies the vision of a sustainable future. Yet, this endeavor necessitates a comprehensive approach to safety. Rooftop installations often involve working at heights, demanding strict adherence to safety protocols.

Personal protective equipment (PPE) takes center stage in rooftop installations. The use of harnesses, helmets, and non-slip footwear is not a mere suggestion; it's a requirement. These gears form a protective barrier against potential falls, ensuring that installers remain secure and confident as they navigate the rooftop terrain.

Accessing the rooftop is a pivotal moment, and it should be executed with precision. The ladder or scaffold system employed should be stable and reliable. Moreover, maintaining proper weight distribution across the roof's surface is essential to prevent structural damage. Windy or rainy weather conditions add complexity to rooftop installations, demanding a pause until the weather becomes conducive to safe work.

Ground Installations

The ground beneath our feet presents a different canvas for solar installations. Ground-mounted systems have their unique set of safety considerations, primarily revolving around heavy machinery and site preparation.

Prior to installation, thorough site preparation is indispensable. This involves clearing the installation site of any obstacles, ensuring stable ground conditions, and marking underground utilities. Such meticulous groundwork minimizes the risk of accidents and disruptions during the installation process.

Heavy machinery, often used in ground installations, requires operators with specialized training. Professionals well-versed in the mechanics of these machines ensure safe and efficient operation. Implementing clear communication channels and maintaining a safe distance between personnel and machinery are central to preventing accidents.

Adhering to Local Building Codes and Permit Requirements

In the dynamic landscape of renewable energy, it's not just the physical components of solar installations that matter; it's also about seamless integration within the regulatory framework of a locality. Adhering to local building codes and securing the necessary permits ensures a legally compliant installation and a safe and efficient one.

Navigating Building Codes

Building codes serve as the compass guiding solar installations toward safety and efficacy. These codes encompass a wide spectrum of requirements, from structural integrity to electrical safety standards. Ignoring or underestimating the importance of building codes can lead to subpar installations fraught with safety hazards.

Understanding the building codes relevant to the installation area is an essential prelude. This understanding goes beyond a superficial grasp; it involves a comprehensive knowledge of the codes' intricacies and implications. It's about grasping how these codes shape the installation, influencing everything from the type of materials used to the layout of the components.

Obtaining Permits

Permits are the legal gateways that authorize the installation of solar power systems. The process of obtaining permits varies based on factors such as system size, location, and complexity. It's a process that demands meticulous attention to detail and timely submission of necessary documents.

Engaging with local permitting authorities is an invaluable strategy. It establishes a dialogue, allowing installers to understand the specific requirements and expectations of the authorities. Moreover, it paves the way for a smoother and more efficient permitting process, eliminating potential roadblocks that could delay the installation timeline.

CHAPTER 2: INSTALLING SOLAR PANELS ON DIFFERENT SURFACES (ROOFS, GROUND, ETC.)

In the realm of sustainable energy solutions, solar panels have emerged as a revolutionary technology that offers both environmental benefits and economic advantages. These photovoltaic (PV) systems convert sunlight into electricity, reducing dependency on fossil fuels and mitigating the impact of climate change. However, the effectiveness of solar panels depends not only on their quality and capacity but also on their strategic placement. In this chapter, we delve into the diverse methods of solar panel installation on various surfaces, ranging from roofs to the ground, and even explore the realm of portable solar arrays.

Roof-Mounted Solar Panel Installation: Pitched and Flat Roofs

Pitched Roofs

Pitched roofs, characterized by their steep incline, present a distinctive challenge and opportunity for solar panel installation. This method is especially prevalent in residential properties, where the roof's angle can be leveraged for optimal sun exposure. The installation process for pitched roofs requires a meticulous approach to ensure both the roof's structural integrity and the solar panels' maximum efficiency.

Assessment and Planning

The journey begins with a comprehensive assessment of the roof's condition and suitability for solar panel installation. Factors such as the roof's material, age, load-bearing capacity, and orientation towards the sun are carefully evaluated. A site visit also identifies any potential sources of shading, such as nearby trees or buildings, which can impact the efficiency of the solar panels.

Mounting System Selection

Selecting the appropriate mounting system is crucial for the success of the installation. The chosen system should be compatible with the roof's material, pitch, and structural characteristics. Common mounting options include roof penetrations, ballasted systems, and hybrid solutions that combine various techniques.

Panel Placement and Orientation

Achieving maximum energy production hinges on the correct placement and orientation of the solar panels. Panels are positioned to minimize shading and maximize exposure to sunlight throughout the day. Depending on the specific latitude of the installation site, panels are angled to capture the most sunlight during peak hours.

Wiring and Inverter Installation

The electrical components of the solar panel system are equally vital. Wiring is routed meticulously, following safe and efficient paths to connect the panels and ensure an uninterrupted flow of electricity. An inverter, the heart of the system, is installed to convert the direct current (DC) produced by the panels into usable alternating current (AC) electricity.

Final Checks and Quality Assurance

Once the installation is complete, a thorough inspection is carried out to ensure that the system is structurally secure and functioning optimally. The integrity of the roof is a top priority, and any potential leaks or weak points are addressed before concluding the installation.

Flat Roofs:

Flat roofs, often found in commercial and industrial settings, introduce different challenges due to their lack of natural inclination. However, they offer distinct advantages, such as larger surface areas and the ability to customize panel orientation. The installation process for flat roofs requires careful planning, engineering expertise, and a focus on safety.

Structural Evaluation:

The first step in installing solar panels on a flat roof involves a thorough structural evaluation. Engineers assess the roof's load-bearing capacity to accommodate the panels' additional weight, mounting systems, and potential snow loads.

Mounting and Ballasting Solutions:

Unlike pitched roofs, flat roofs require mounting systems that provide an appropriate tilt angle for the panels. These systems may involve tilting the panels towards the sun at a calculated angle to maximize energy capture. To prevent panels from being dislodged by strong winds, ballast, such as weighted blocks or containers, is employed to secure the system.

Panel Layout for Efficiency:

The arrangement of panels on a flat roof is critical to avoid shading and ensure efficient airflow. Proper spacing between panels is maintained to prevent overheating and optimize energy production.

Electrical Integration:

As with other installation methods, proper wiring and inverter installation are essential for converting and transmitting the generated electricity. Wiring pathways are designed with both functionality and aesthetics in mind, while inverters are strategically placed to minimize energy losses.

Maintenance Accessibility:

Designing access pathways on flat roofs is essential to facilitate routine maintenance, including panel cleaning and system inspections. Regular maintenance ensures the system's longevity and sustained performance.

Ground-Mounted Solar Arrays: Foundation and Tilt Angles

Site Selection and Preparation

Ground-mounted solar arrays offer a versatile solution for installations where roof space is limited or specific sun exposure angles are desired. The installation process begins with selecting a suitable site and preparing it for the solar array.

A thorough site assessment involves evaluating factors such as solar irradiance, shading patterns, soil conditions, and accessibility. The solar array's orientation and tilt angles are determined based on the latitude of the site to optimize energy capture.

The foundation is a critical component of ground-mounted installations. Engineers choose the appropriate foundation type, which may involve concrete footings, ground screws, or ballasts, based on the soil composition and load-bearing capacity.

Mounting Structure and Tilt Angle Optimization:

The mounting structure forms the backbone of the ground-mounted solar array. This structure holds the panels and determines their tilt angle for maximum sun exposure.

Ground-mounted systems typically use rack or frame structures made of durable materials like steel or aluminum. These structures provide stability and support for the solar panels.

The tilt angle of the panels is calculated based on the latitude of the installation site. The goal is to position the panels at an angle that allows them to capture the most sunlight throughout the year. This optimization process requires precise calculations to achieve optimal energy production.

Wiring, Inverter Setup, and Maintenance:

Just like other installation methods, proper wiring, inverter setup, and maintenance are integral to ensuring the efficiency and longevity of the ground-mounted solar array.

Wiring is carefully routed and connected to create a reliable network that allows the electricity generated by the panels to flow to the inverter and, ultimately, to the electrical grid or on-site usage.

The inverter setup remains consistent with other installation methods. The inverter converts the DC electricity produced by the panels into AC electricity, which is suitable for consumption and grid integration.

Ground-mounted systems offer advantages in terms of accessibility for maintenance and cleaning. Regular cleaning of the panels to remove dust, debris, and bird droppings is essential to maintain their performance. In addition, routine system inspections ensure that all components are functioning optimally.

Portable Solar Panels: Setting Up Solar Arrays Anywhere

Design and Portability

Portable solar panels introduce a dynamic and versatile dimension to solar energy utilization. These panels are meticulously designed to be lightweight, compact, and easily transportable.

Portable panels often feature foldable or rollable designs that facilitate transportation and storage. These designs ensure that users can carry the panels conveniently to various locations.

Portable solar systems come equipped with built-in charge controllers and batteries. The charge controller regulates the flow of electricity from the panels to the batteries, preventing overcharging. Batteries store the generated energy, making it available for use when needed.

Deployment and Use

Setting up a portable solar array is a straightforward process that empowers users to harness solar power wherever they go.

Users unfold or unroll the panels in an area with ample sunlight. Positioning the panels at an optimal angle ensures effective sunlight capture and energy production.

Portable solar panels have the capability to directly charge electronic devices like smartphones, tablets, laptops, and even portable power banks. Users connect their devices to the panel's charging ports, converting solar energy into usable power.

Versatility and Maintenance

The versatility of portable solar panels opens up a multitude of applications and scenarios where solar energy can be utilized.

Portable panels find utility in a diverse range of situations, including camping trips, hiking expeditions, RV adventures, boating trips, emergency power generation during natural disasters, and remote off-grid locations.

While portable solar panels require less maintenance compared to fixed installations, regular cleaning is essential to ensure optimal performance. Simple cleaning routines, coupled with proper storage in a dry and secure environment, prolong these systems' lifespan.

CHAPTER 3: CONNECTING BATTERIES AND CONFIGURING ENERGY STORAGE

In the quest for sustainable energy solutions, the integration of batteries into our power systems has become increasingly essential. Whether it's for residential solar setups, industrial backup power, or electric vehicles, the efficient connection and configuration of batteries play a pivotal role in optimizing energy storage systems. This chapter delves into the intricate details of battery bank configuration, wiring techniques for safety and efficiency, and the critical aspects of battery maintenance and capacity testing.

Battery Bank Configuration: Series vs. Parallel Connections

The architecture of an energy storage system lies at the core of its functionality, and the configuration of battery banks is pivotal in determining its performance. Two primary strategies stand out in this regard: series and parallel connections. Each presents its own set of advantages and considerations, requiring careful evaluation based on the system's intended purpose.

Series Connections

Series connections entail linking batteries end to end, effectively increasing the cumulative voltage of the battery bank while maintaining a constant current flow. The potential use cases for this configuration are diverse, ranging from applications that necessitate high voltage levels, such as grid-connected installations and electric vehicles.

While series connections offer benefits like elevated voltage output, they also introduce certain complexities. The uniformity of the charging and discharging process across all batteries becomes crucial. Discrepancies in battery performance within a series can lead to imbalances, potentially compromising the overall effectiveness of the energy storage system.

Parallel Connections

Parallel connections, in contrast, involve linking batteries' positive terminals together and their negative terminals likewise. This arrangement maintains the bank's voltage while enhancing its overall current capacity. Such a configuration finds its utility in scenarios requiring heightened current output, such as off-grid setups and high-power applications.

However, implementing parallel connections requires meticulous attention to detail. Batteries, even of the same model, can exhibit slight differences in terms of their internal resistance and capacity. Ensuring an equitable distribution of the charging and discharging processes is crucial to avoid imbalances that could undermine the system's efficiency.

Combined Connections

In real-world applications, the choice often veers toward a hybrid approach that combines both series and parallel connections. This offers a balanced solution where the system can achieve the desired voltage levels while accommodating the required current capacity.

The decision between series, parallel, or hybrid connections hinges on a multitude of factors: the specific purpose of the energy storage system, available space for the installation, budget considerations, and the sought-after efficiency levels. Therefore, attaining a comprehensive grasp of these configurations is essential for devising energy storage systems that align harmoniously with the designated performance benchmarks.

Proper Wiring and Fusing for Battery Safety and Efficiency

The intricate web of wires and the judicious implementation of fusing within an energy storage system are pivotal factors that can spell the difference between smooth operation and potential hazards. Proper wiring and fusing practices ensure that the system functions optimally while maintaining safety standards.

Wiring Considerations: When it comes to wiring, several key considerations come into play:

1. **Wire Gauge:** The gauge or thickness of the wire plays a critical role in determining the safe maximum current it can handle. Selecting an appropriate wire gauge is essential to prevent overheating and voltage drop. Opting for wires that are too thick can lead to increased costs and unnecessary consumption of space.

2. **Cable Length:** Longer cables introduce heightened resistance, culminating in energy losses manifested as heat. The strategy here is to minimize cable length and establish meticulous routing to mitigate these losses.

3. **Connection Type:** Ensuring reliable connectivity between components involves employing proper connection methods, such as crimping or soldering. Loose connections can trigger voltage drops, reduced efficiency, and, at worst, arc faults that could result in accidents.

4. **Isolation and Protection:** To preempt accidental contact and short circuits, adequate insulation of wires is paramount. Moreover, safeguarding the wiring against physical damage and environmental elements is crucial to sustaining optimal functioning.

Fusing for Safety: The inclusion of fuses in battery systems is non-negotiable from a safety perspective. Fuses serve as sacrificial components that rupture the circuit if the current surpasses safe thresholds, mitigating potential risks like overheating and fire.

Selecting the Right Fuse: Fuses should be selected based on the maximum current capacity of the circuit. An optimal fuse rating ensures that the system is protected without triggering unnecessary tripping during normal operation.

Placement: Strategic placement of fuses within the circuit is imperative. Positioning fuses close to the battery bank and at critical junctions in the wiring curtails the unprotected circuit length, consequently reducing potential damage in the event of a fault.

Fuse Types: Two main types of fuses are available: fast-acting and slow-blow. Fast-acting fuses respond promptly to overcurrent scenarios, making them suitable for safeguarding sensitive components. Slow-blow fuses, on the other hand, are designed to withstand temporary overloads, making them a viable option for circuits where inrush currents are common.

Backup Protection: Complementary protective elements like circuit breakers can be deployed alongside fuses to enhance safety. These devices can be manually reset following the clearance of a fault, adding an extra layer of protection.

Battery Maintenance and Capacity Testing

Ensuring battery banks' sustained health and efficacy necessitates the conscientious adoption of meticulous maintenance practices and regular capacity testing.

Maintenance Practices:

1. **Visual Inspection:** Conducting frequent visual inspections is a first line of defense against potential issues. Corrosion, leaks, swelling, or physical damage are telltale signs that necessitate swift intervention to maintain battery health.
2. **Temperature Control:** Batteries function optimally within specific temperature ranges. Implementing temperature controls, including ventilation and cooling systems, helps forestall overheating, extending the operational lifespan of the batteries.
3. **Cleanliness:** The cleanliness of batteries is paramount. Dust and debris accumulation raises the risk of short circuits and impedes heat dissipation, contributing to reduced performance and self-discharge.
4. **Equalization Charging:** For systems featuring multiple batteries, periodic equalization charging is crucial. This process serves to balance the charge levels of individual batteries, promoting uniform performance across the entire battery bank.

Capacity Testing:

1. **Regular Testing:** Over the course of time, batteries can witness a decline in capacity due to chemical changes. Regular capacity testing, typically an annual practice, involves evaluating the actual capacity against the manufacturer's specifications.

2. **Discharge Testing:** During capacity testing, batteries are fully charged and subsequently discharged in a controlled manner. This process enables a precise assessment of the battery's ability to store and release energy effectively.

3. **Interpretation:** The results of these tests are then juxtaposed against the initial capacity and the manufacturer's stipulated values. Substantial drops in capacity may indicate the need for battery replacement or necessitate adjustments in the charging regimen.

4. **Record Keeping:** Meticulous documentation of capacity test outcomes serves as a historical record that tracks the battery's performance trajectory. Such records facilitate informed decisions concerning proactive maintenance measures.

Battery banks' harmonious connection and configuration constitute the bedrock of optimized energy storage systems. A discerning comprehension of series and parallel connections equips system designers to tailor solutions to meet specific requisites. The proper execution of wiring techniques and the prudent integration of fusing guarantee safety and efficiency, curtailing the potential for accidents and operational disruptions. Lastly, the conscientious adoption of rigorous maintenance practices and periodic capacity testing secures the endurance and efficacy of batteries, fostering reliable and efficient energy storage solutions that contribute tangibly to a sustainable energy future.

CHAPTER 4: WIRING YOUR SOLAR SYSTEM: STEP-BY-STEP GUIDE

The journey towards harnessing solar energy involves not only the installation of solar panels but also the intricate process of wiring the entire solar system. Proper wiring is the backbone of a functional and safe solar setup, ensuring that the energy generated is efficiently collected, stored, and distributed.

Electrical Circuit Design and Wiring Diagrams

Before any physical wiring takes place, careful planning and circuit design are essential. A comprehensive understanding of the components in your solar system and how they interact will guide you in creating an effective and optimized wiring layout.

Planning Your Electrical Circuit

Before diving into the intricacies of wiring diagrams, it's essential to outline your electrical circuit design. This involves mapping out how energy will flow from the solar panels to the charge controllers, batteries, inverters, and finally to your appliances. Start by categorizing your loads based on their voltage requirements, such as DC loads (lights, fans) and AC loads (appliances, electronics). This categorization influences how your components will be connected.

Choosing the Right Wire Sizes

Selecting the appropriate wire sizes is critical for minimizing energy losses, preventing overheating, and ensuring the safety of your system. The wire size should be determined based on the maximum current your system will carry and the distance the wire will span. Refer to wire sizing charts that consider both current capacity and voltage drop. Larger wire gauges reduce resistance and energy losses, particularly for longer cable runs.

Creating Wiring Diagrams

Wiring diagrams are visual representations of your electrical circuit, illustrating how components are interconnected. These diagrams provide a clear overview of your system's layout, helping you troubleshoot issues and make informed decisions during installation and maintenance. Start by sketching a diagram on paper, indicating the locations of solar panels, charge controllers, batteries, inverters, and loads. Then translate your hand-drawn sketch into a digital format using software tools like AutoCAD or specialized online diagram creators.

Differentiating AC and DC Wiring

In off-grid systems, you'll encounter both AC and DC wiring. AC wiring carries the energy produced by your inverter to power AC appliances, while DC wiring connects the solar panels, batteries, and DC loads. It's crucial to keep these two types of wiring separate and properly labeled to prevent confusion and ensure safety during installation and maintenance.

Color Coding and Labeling

To simplify troubleshooting and maintenance, color coding and labeling are invaluable. Follow industry-standard color codes for both AC and DC wiring. For instance, red or black wires are often used for positive DC connections, while white or gray wires represent neutral or negative connections. Properly labeling components, junction boxes, and breaker panels allows you to quickly identify and address any issues that may arise.

Safety should always be a priority when working with electrical circuits. Ensure that your wiring meets the electrical codes and regulations of your region. Proper grounding, installing circuit breakers, and using safety disconnects are critical for preventing electrical hazards.

As you finalize your wiring diagram, don't forget to document your system comprehensively. This documentation should include a detailed list of components, wire gauges, breaker sizes, and other pertinent information. This documentation becomes a valuable reference for future maintenance, troubleshooting, or system upgrades.

Wiring Solar Panels to Charge Controllers and Batteries

As you delve deeper into the practical aspects of your off-grid solar power system, understanding how to wire solar panels to charge controllers and batteries becomes a crucial skill. Properly connecting these components ensures the smooth flow of energy and the efficient charging of your batteries. This chapter guides you through the steps of this essential process, from connecting solar panels to charge controllers to linking your batteries to the energy source.

Connecting Solar Panels to Charge Controllers

1. Before you begin, ensure you have the required components, including solar panels, charge controllers, solar cables, and appropriate connectors.

2. Each solar panel has a positive and a negative lead. Usually, red signifies positive and black indicates negative. Confirm the polarity markings on both the panels and the charge controller.

3. Depending on your system's voltage and current requirements, you can connect solar panels in parallel (positive to positive, negative to negative) for higher current output or in series (positive to negative) for increased voltage. Check your charge controller's specifications to determine the suitable configuration.

4. Choose appropriately sized solar cables to connect panels to the charge controller. Oversized cables minimize energy losses due to resistance. Use MC4 connectors or other compatible connectors designed for outdoor use.

5. Connect the positive lead of the solar panel to the positive input terminal of the charge controller and the negative lead to the negative terminal. Make sure connections are secure and properly tightened.

6. Ensure that the voltage and current produced by your solar panels match the specifications of the charge controller. A mismatch can damage the charge controller and batteries.

Wiring Batteries to Charge Controllers

1. Ensure you have the necessary batteries, battery cables, charge controller, and appropriate fuses.

2. Off-grid systems often use multiple batteries connected in series to create a battery bank. The combined voltage of the batteries determines the overall system voltage.

3. Connect the positive terminal of one battery to the negative terminal of the adjacent battery using battery cables. Repeat this process until you've linked all the batteries in series.

4. Attach the positive cable from the battery bank to the charge controller's battery terminal. Connect the negative cable to the charge controller's negative terminal.

5. Install fuses or circuit breakers between the battery bank and the charge controller for added protection against overcurrent or short circuits.

6. Configure the charge controller according to the battery bank's specifications, including voltage settings, charging profiles, and low-voltage disconnect levels.

AC and DC Wiring for Power Distribution in Off-Grid Systems

In off-grid solar systems, power distribution involves direct current (DC) and alternating (AC) wiring. This part of the chapter dives into the complexities of safely distributing the harnessed solar energy.

DC Wiring

DC wiring connects various components of your solar setup that operate on direct current. This includes connecting the batteries, charge controller, and DC loads. Use appropriately sized and rated cables to minimize energy loss. Implement circuit protection measures like fuses or circuit breakers.

Inverter Installation

An inverter is employed to convert DC power to usable AC power (for standard household appliances). Proper installation of the inverter is crucial for efficient energy conversion. Connect the inverter to the batteries following the manufacturer's instructions.

AC Wiring

AC wiring handles the distribution of converted power to your household appliances. Install circuit breakers and outlets according to local electrical codes. Keep AC and DC wiring physically separated to prevent accidents.

Grounding and Safety

Proper grounding is imperative in any electrical system. Ground the AC and DC sides of your solar system to prevent electric shock hazards. Implement surge protection devices to safeguard against power surges caused by lightning strikes or other anomalies.

The process of wiring your solar system demands meticulous planning, attention to detail, and adherence to safety protocols. A well-designed and properly wired solar setup ensures optimal energy collection, storage, and distribution. From understanding the basics of load analysis to mastering the art of AC and DC wiring, this chapter has unveiled the intricate steps involved in wiring a solar system. As you embark on this journey, remember that a successfully wired solar system contributes to sustainable energy usage and guarantees the safety of all who interact with it.

BOOK 5

OFF-GRID POWER MANAGEMENT AND OPTIMIZATION

CHAPTER 1: UNDERSTANDING ENERGY LOAD MANAGEMENT

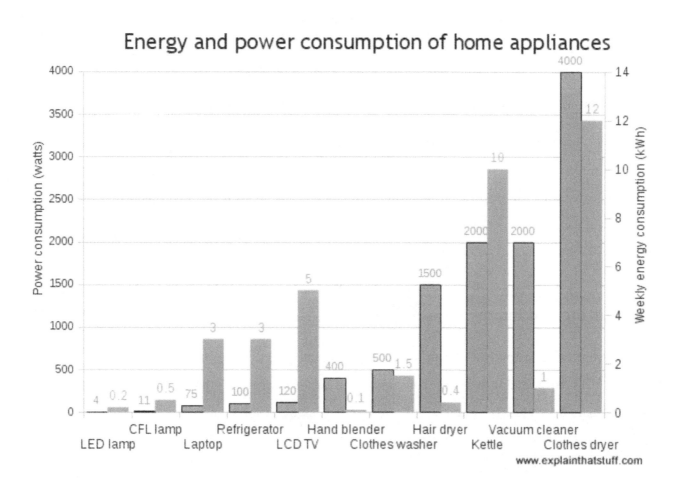

Energy and power consumption of home appliances

www.explainthatstuff.com

Energy consumption has become an integral part of modern life. The increasing reliance on electronic devices, appliances, and technologies has led to a surge in energy demand. This surge not only impacts our monthly bills but also significantly strains the environment. In response to these challenges, the concept of energy load management has gained prominence as an effective strategy to optimize energy consumption. This chapter delves into the intricacies of energy load management, encompassing energy load analysis, load shifting and conservation strategies, and the crucial aspect of load priority in managing essential versus non-essential loads.

Energy Load Analysis: Identifying High-Consumption Devices

Energy load analysis serves as the foundation of effective load management. Before implementing any energy-saving strategy, it is imperative to have a comprehensive understanding of which devices contribute most significantly to the overall energy consumption. This analysis involves the identification of high-consumption devices, which are often the culprits behind inflated energy bills.

It is recommended to employ energy monitoring tools to conduct a meaningful energy load analysis. These tools provide real-time data on the energy consumption of individual devices. Smart energy meters, for instance, offer insights into how much energy each device is using, allowing homeowners and businesses to pinpoint the sources of high consumption. By analyzing this data, patterns of usage can be identified. Appliances that consistently draw substantial power, such as air conditioners, water heaters, and refrigerators, come under scrutiny.

Energy load analysis isn't limited to residential spaces alone. Industries and commercial establishments can also benefit from this practice. Large machinery, lighting systems, and HVAC (Heating, Ventilation, and Air Conditioning) systems can contribute significantly to energy loads in such settings. Employing sub-metering in industrial and commercial setups helps isolate energy usage patterns for various operations, making it easier to allocate resources more efficiently.

Load Shifting and Energy Conservation Strategies

Once high-consumption devices are identified, load-shifting and energy conservation strategies come into play. Load shifting involves redistributing the energy consumption of certain devices to off-peak hours when electricity demand is lower. This approach reduces strain on the grid during peak times and often comes with lower electricity rates, incentivizing consumers to shift their usage patterns.

One of the most common examples of load shifting is setting appliances such as washing machines, dishwashers, and pool pumps to operate during off-peak hours. This approach doesn't necessarily reduce the total energy consumed by these devices, but it ensures that they draw power when the grid can handle it more efficiently.

On the other hand, energy conservation strategies aim to reduce devices' actual energy consumption. This can be achieved through various means, such as upgrading to more energy-efficient appliances, using LED lighting, and improving insulation to reduce the workload on heating and cooling systems.

For instance, the adoption of smart thermostats can significantly contribute to energy conservation. These devices learn the occupants' schedules and preferences, adjusting the temperature accordingly. When no one is home, the thermostat can automatically set the temperature higher in the summer and lower in the winter, reducing the workload on HVAC systems.

Renewable energy sources also play a vital role in energy conservation. Solar panels, for instance, generate clean energy from the sun, which can offset a significant portion of traditional energy consumption. By integrating these systems into load management strategies, consumers can reduce their reliance on the grid and potentially even sell excess energy back to utility companies.

Load Priority and Managing Essential vs. Non-Essential Loads

Load priority is a crucial consideration in energy load management. Not all devices and appliances are of equal importance, and during periods of high demand or energy scarcity, tough decisions may need to be made regarding which loads to prioritize.

Essential loads typically include devices that are critical for health, safety, or business continuity. Medical equipment, security systems, and essential lighting fall into this category. Non-essential loads, on the other hand, encompass devices that, while convenient, are not critical for immediate well-being or operations. These might include entertainment systems, decorative lighting, or certain appliances.

During peak demand times or energy shortages, load shedding can be implemented. Load shedding involves deliberately disconnecting non-essential loads temporarily to reduce the overall demand on the grid. This proactive approach prevents blackouts or brownouts by ensuring that critical loads have enough power. Advanced load management systems often use a combination of load shifting and load shedding to optimize energy consumption. Smart systems can automatically adjust loads based on real-time demand and availability, ensuring that essential loads are always prioritized while non-essential loads are adjusted or shifted to balance the overall energy consumption.

Energy load management is a multifaceted approach to optimizing energy consumption. Through energy load analysis, the identification of high-consumption devices sets the stage for load shifting and energy conservation strategies. These strategies, in turn, contribute to reduced energy bills and a lighter environmental footprint. However, the crux of successful energy load management lies in understanding load priority – effectively managing essential and non-essential loads during times of high demand or scarcity. As technology continues to advance, so do the opportunities for more sophisticated and automated load management systems, promising a future where energy is used more intelligently and sustainably.

CHAPTER 2: ENERGY EFFICIENCY TIPS FOR TINY LIVING SPACES

In an age of environmental consciousness and rapidly depleting resources, the concept of sustainable living has taken center stage. With the rise of minimalistic lifestyles and the ever-growing trend of tiny homes, the need for energy efficiency in compact living spaces has never been more crucial. This chapter delves into energy efficiency tips tailored to the unique challenges and opportunities of tiny living spaces.

Energy-Efficient Appliances and Lighting Solutions

The significance of each appliance and lighting fixture is amplified in the realm of tiny living spaces, where square footage is a precious commodity. The selection of energy-efficient appliances and smart lighting solutions can profoundly impact the overall energy consumption and sustainability of such compact living environments.

When it comes to appliances, the rule of thumb is to choose quality over quantity. In the limited space available, each appliance needs to earn its place by serving a purpose and doing so with minimal energy consumption. Energy Star ratings emerged as a boon for environmentally conscious consumers. These ratings clearly indicate a product's energy efficiency, allowing consumers to make informed choices. When outfitting a tiny living space, appliances such as refrigerators, ovens, and washing machines should be scrutinized for their Energy Star labels.

The refrigerator, often a power-hungry appliance, presents a prime energy-saving opportunity. Opt for a compact, Energy Star-rated refrigerator that aligns with the needs of your lifestyle. Consider the layout of the fridge as well; side-by-side models tend to be less efficient due to the larger surface area exposed when opened. A top or bottom freezer configuration is generally more energy-efficient.

While indispensable for cooking, ovens can also be a source of energy wastage. Convection ovens, which circulate hot air around the food, can cook dishes more quickly and at lower temperatures than traditional ovens. This can lead to significant energy savings over time. Additionally, investing in a microwave or toaster oven for smaller cooking tasks can help mitigate the energy demands of a larger oven.

Washing machines are another area where energy efficiency should be a priority. Front-loading washing machines are typically more energy-efficient than top-loading models. They use less water and require less energy to heat that water, resulting in both water and energy savings. Furthermore, always opt for full loads to maximize the efficiency of each laundry cycle.

Lighting is a vital component of tiny living spaces. The confined dimensions of such homes often mean limited natural light, necessitating more reliance on artificial lighting sources. However, this offers an opportunity to optimize energy consumption through strategic lighting choices.

The traditional incandescent bulb, while inexpensive upfront, is highly inefficient in the long run. It emits a significant portion of its energy as heat rather than light. Compact fluorescent lamps (CFLs) and light-emitting diode (LED) bulbs come to the rescue. Not only do these bulbs consume a fraction of the energy of incandescent bulbs, but they also have significantly longer lifespans, reducing the frequency of replacements. While LED bulbs are generally more expensive, their durability and energy savings justify the initial investment.

In the context of tiny living, lighting should be energy-efficient and adaptable to different activities and moods. Smart lighting solutions provide the answer. These systems allow users to remotely control the brightness and color of the light, offering customization and flexibility. With the ability to program lighting schedules and even sync with circadian rhythms, smart lighting ensures that energy is only expended when needed.

Insulation and Passive Heating/Cooling Techniques

Regarding energy efficiency in tiny living spaces, the concept of "less is more" extends beyond just possessions – it also encompasses energy consumption. The limited area of these compact homes demands a strategic approach to maintaining a comfortable indoor environment without relying heavily on energy-consuming heating and cooling systems. This is where insulation and passive techniques come into play as crucial components of sustainable living.

Proper insulation acts as a thermal barrier, preventing the exchange of heat between the interior and exterior environments. In tiny homes, where space is at a premium, insulation serves as a key defense against extreme temperatures. High-quality insulation materials strategically placed in walls, floors, and ceilings can make a significant difference in the energy requirements of the space.

Closed-cell spray foam insulation is a prime example of a highly effective insulating material. Not only does it provide excellent thermal resistance, but it also acts as an airtight sealant, minimizing drafts and heat leakage. This type of insulation is particularly beneficial in small spaces, where even a small gap can noticeably impact temperature regulation.

However, insulation is only part of the equation. To truly achieve energy efficiency, passive heating and cooling techniques need to be incorporated into the design and construction of the tiny home. Passive techniques utilize the natural elements – sunlight, airflow, and the thermal mass of materials – to regulate indoor temperatures.

In colder months, passive heating can significantly reduce the need for external heating sources. The strategic placement of windows is a critical factor in this regard. South-facing windows capture the maximum amount of sunlight throughout the day. Using energy-efficient windows with low emissivity coatings maximizes the heat gain while minimizing heat loss.

Thermal mass materials, such as stone or concrete, play a pivotal role in passive heating. These materials have the ability to absorb and store heat during the day when sunlight is plentiful. As the temperature drops in the evening, they release this stored heat, providing a gradual and consistent source of warmth. Integrating thermal mass into the tiny home's design, such as through flooring or interior walls, ensures that the heat absorbed during the day is effectively utilized to maintain comfort during cooler hours.

Conversely, passive cooling techniques are indispensable in warmer months. Cross-ventilation, facilitated by strategically placed windows and vents, allows fresh air circulation, carrying heat away from the living space. Designing the home with prevailing winds in mind further enhances this effect.

Reflective roofing materials offer another means of passive cooling. By reflecting a significant portion of the incoming solar radiation, these materials prevent excess heat absorption, thereby reducing the overall indoor temperature. Combined with proper insulation, reflective roofing can contribute to a naturally cooler interior environment.

Shading elements, such as pergolas, awnings, or deciduous trees, provide valuable protection from the sun's rays. By strategically placing these elements on the sun-exposed sides of the tiny home, you can prevent direct sunlight from entering, thereby reducing the need for active cooling systems.

Insulation and passive heating/cooling techniques form a dynamic duo in the quest for energy efficiency in tiny living spaces. Proper insulation sets the foundation by minimizing heat transfer, while passive techniques harness the power of nature to maintain optimal indoor temperatures. The synergy between these approaches reduces energy consumption and enhances the sustainability and livability of these compact homes.

Smart Home Automation for Optimal Energy Usage

The advent of the Internet of Things (IoT) has revolutionized the way we interact with and manage our living spaces in the context of tiny homes, where every inch and watt counts. Smart home automation emerges as a formidable tool in the pursuit of energy efficiency and sustainability.

At its core, smart home automation entails the integration of various devices and systems into a centralized network that can be controlled and monitored remotely. This networked approach enables homeowners to gain unprecedented insights into their energy usage patterns and exercise precise control over their living environment.

The smart thermostat is a cornerstone of smart home automation in the context of energy efficiency. Armed with advanced sensors and algorithms, these devices learn the occupants' routines and preferences. This knowledge is then used to optimize temperature settings, reduce energy waste when the home is unoccupied, or adjust conditions for comfort before occupants return.

The benefits of smart thermostats extend beyond basic scheduling. Some models can integrate weather forecasts into their algorithms, anticipating temperature fluctuations and adjusting settings accordingly. Others can be controlled remotely via smartphone apps, allowing users to modify settings even when they're away from home. These capabilities ensure that energy is expended only when necessary without compromising comfort.

Energy monitoring systems, another facet of smart home automation, provide real-time insights into energy consumption patterns. These systems collect data from various appliances and devices, giving homeowners a detailed breakdown of where and when energy is being used. Armed with this information, residents can identify energy-hungry appliances and adjust their usage accordingly.

Advanced energy monitoring systems can even predict usage trends based on historical data. This forecasting capability empowers homeowners to plan their energy consumption more effectively. For instance, if a certain appliance tends to consume more energy during peak hours, the system can suggest running it during off-peak periods to reduce costs.

Smart plugs and power strips add a practical layer of control to the energy management arsenal. These devices, which can be integrated into the home automation network, allow users to remotely turn off electronics and appliances that might otherwise continue drawing energy in standby mode. They can also be programmed to follow a schedule, ensuring that devices are powered down when not in use.

CHAPTER 3: OPTIMIZING POWER CONSUMPTION IN CABINS AND RVS

As the world continues to navigate an era of environmental consciousness and sustainable living, the optimization of power consumption in various domains has become a paramount concern. Among these, the domains of cabins and recreational vehicles (RVs) stand out prominently. Cabins, whether they serve as primary residences or vacation getaways, and RVs, beloved by nomads and adventure seekers, share a common challenge: balancing comfort and power efficiency. This chapter delves into the multifaceted realm of optimizing power consumption in cabins and RVs, exploring energy-saving cabin designs, insulation techniques, RV energy management at off-grid and on-grid campsites, and sustainable water and waste management solutions.

Energy-Saving Cabin Design and Insulation

In the pursuit of energy-efficient cabin design, the intricate interplay between architectural ingenuity and sustainable principles takes center stage. A cabin is more than just a structure; it combines nature and human habitation harmoniously. Energy-saving cabin design encapsulates a holistic approach that minimizes energy consumption and accentuates the connection between indoor and outdoor spaces.

Passive solar design, a cornerstone of energy-efficient architecture, leverages the sun's path to harness or repel heat as needed. When applied to cabins, this concept is masterfully orchestrated. South-facing windows capture the low winter sun, flooding the interior with warmth and light. Adequately sized overhangs provide shade during the summer months when the sun's angle is higher, preventing excessive heat buildup. This dance between architectural elements and the sun's trajectory ensures a natural equilibrium within the cabin.

In addition to solar orientation, the choice of construction materials is a pivotal determinant of energy efficiency. Insulated concrete forms (ICFs), known for their excellent insulation properties and structural stability, form a formidable barrier against external temperature fluctuations. They encapsulate a concrete core, effectively buffering the indoor environment from the vagaries of weather.

Structural insulated panels (SIPs), comprising a foam core sandwiched between oriented strand board (OSB) panels, achieve impressive insulation levels while reducing thermal bridging. This innovative construction method creates a tightly sealed envelope, minimizing heat loss or gain. Moreover, the use of sustainable, locally sourced materials aligns with the ethos of responsible construction, further reducing the cabin's environmental impact.

As much as the exterior matters, a well-insulated cabin envelope is equally crucial. Spray foam insulation, a versatile option, fills every nook and cranny and seals off potential air leaks. Its expansive nature ensures that the cabin remains airtight, thwarting drafts that could compromise energy efficiency. Fiberglass and cellulose insulation, while more traditional, still hold their own when properly installed. They provide commendable resistance to heat transfer, bolstering the cabin's ability to maintain a consistent indoor temperature.

Windows, often regarded as a cabin's eyes to the world, play a substantial role in energy-saving efforts. High-performance windows with low-emissivity coatings and multiple panes possess exceptional thermal insulation properties. They act as a selective barrier, allowing natural light to filter in while thwarting heat infiltration during summer and heat loss during winter.

RV Energy Management: Off-Grid vs. On-Grid Campsites

The nomadic allure of RV living conjures images of open roads and boundless exploration. Yet, this freedom comes with a trade-off: limited energy resources. RV energy management isn't just a practical consideration; it's a lifestyle commitment to sustainable mobility.

Off-grid RV campsites, often nestled in remote wilderness, beckon with the promise of untamed beauty. Here, self-sufficiency becomes a way of life. Solar panels, akin to miniature power plants, adorn the roofs of RVs, silently converting sunlight into electrical energy. This ingenious integration of technology with nature epitomizes eco-friendliness. These panels charge the RV's battery bank, storing energy for when the sun retreats beyond the horizon. Lithium-ion batteries, with their high energy density and longer lifespan, form the heartbeat of the RV's power system, ensuring a consistent supply of lights, appliances, and entertainment.

Thriving in off-grid environments isn't just about energy generation but energy conservation. Energy-efficient appliances, such as compact refrigerators, LED lighting, and propane-fueled stoves, are emblematic of this philosophy. LED lighting, for instance, consumes a fraction of the energy of traditional incandescent bulbs while emitting the same luminosity. Drawing from an onboard tank, propane stoves channel energy directly to cooking, minimizing wastage. The conscious choice to use natural ventilation harnessing cross breezes also lessens reliance on power-hungry air conditioning units.

Contrastingly, on-grid RV campsites offer respite from self-sufficiency. Here, RVs can connect to external power sources, reshaping energy dynamics. While this connection opens access to a steady supply of energy, responsible choices remain crucial. Opting for campgrounds that offer eco-friendly power hookups sourced from renewable energy like solar or wind reflects a commitment to minimizing one's carbon footprint even while plugged in.

However, regardless of the energy source, the tenets of energy-saving persist. Mindful consumption practices endure, manifesting in habits like turning off lights when leaving the RV, optimizing water heater usage, and judiciously using electrical outlets. Such habits, seemingly small, ripple into substantial energy savings over time, aligning seamlessly with the overarching goal of sustainable RV living.

Sustainable Water and Waste Management Solutions

In the realm of cabins and RVs, sustainability stretches beyond energy to encompass water and waste management. These are often overlooked aspects, yet their impact on the environment is profound.

Within cabins, implementing water-saving fixtures initiates a cascade of conservation. Low-flow toilets with innovative flushing mechanisms significantly reduce water consumption without compromising sanitation. Aerated faucets, blending air with water, maintain water pressure while conserving volume. Efficient showerheads, engineered to provide satisfying showers while minimizing water wastage, epitomize the marriage of comfort and sustainability.

Greywater, the residual water from sinks, showers, and washing machines, presents a resource that's often squandered. When incorporated into cabin design, Greywater recycling systems collect and treat this water, transforming it into a valuable resource for irrigation or toilet flushing. Rainwater harvesting systems, another facet of sustainable water management, intercept rainwater from roofs and store it for later use. This harvested rainwater, devoid of chlorine or treatment chemicals, is ideal for non-potable uses like landscape irrigation or toilet flushing.

Responsible water and waste management are equally paramount in the mobile world of RVs. Holding tanks that store wastewater until proper disposal necessitates strategic management. Optimal tank capacity usage minimizes the frequency of emptying, reducing the environmental impact. Choosing campgrounds with dumping stations and adhering to local regulations ensures ethical waste disposal. Embracing biodegradable cleaning products minimizes the introduction of harmful chemicals into wastewater, aligning with the principles of responsible RV living.

Composting toilets offer an elegant solution for those seeking the epitome of sustainability. These self-contained systems convert human waste into nutrient-rich compost, eliminating the need for water-intensive flushing. By mimicking natural decomposition processes, composting toilets symbolize the convergence of innovation and environmental consciousness.

Optimizing power consumption in cabins and RVs transcends mere energy efficiency. It embodies a philosophy of harmonizing with nature, of embracing innovation to tread lightly on the planet. From energy-saving cabin designs and insulation techniques that dance with the sun's path to the conscientious energy management choices made on the open roads of RV adventures and the profound embrace of sustainable water and waste management solutions, these chapters collectively narrate a story of human ingenuity intertwined with environmental stewardship. As the quest for a more sustainable future intensifies, the principles outlined within these chapters serve as guiding stars, illuminating the path to a greener, more harmonious world.

CHAPTER 4: MONITORING AND TROUBLESHOOTING YOUR OFF-GRID SOLAR SYSTEM

In our ever-evolving world, where sustainable energy solutions are becoming increasingly essential, off-grid solar systems have emerged as a beacon of hope. These systems, capable of harnessing the sun's energy and converting it into usable electricity, offer an environmentally friendly and cost-effective alternative to traditional power sources. However, like any complex technological setup, off-grid solar systems require vigilant monitoring and occasional troubleshooting to ensure they operate efficiently and effectively. This chapter delves into the crucial aspects of monitoring, diagnosing issues, and maintaining off-grid solar systems for prolonged and optimal performance.

Remote Monitoring and Data Analysis Tools

The Power of Remote Monitoring

Integrating remote monitoring technology has revolutionized how off-grid solar systems are managed. With remote monitoring, system owners and technicians can access real-time data and insights about the system's performance without being physically present at the installation site. This capability significantly enhances system management efficiency and reduces the need for frequent on-site visits, which can be especially valuable for remote or inaccessible locations.

Data Analysis: A Key to Optimization

Data holds the key to understanding the intricacies of any system's performance, and off-grid solar systems are no exception. Remote monitoring systems continuously gather data points such as energy production, battery levels, and load consumption. Analyzing this data provides valuable insights into the system's patterns, trends, and potential anomalies.

Solar systems generate electricity based on sun exposure, and their efficiency can be affected by factors such as shading, dust accumulation, or module degradation. Data analysis helps detect such issues promptly. For instance, if a sudden drop in energy production is detected, it could indicate a shading issue that needs to be addressed. By identifying such problems early, system owners can take corrective actions before they escalate into more significant complications.

Role of Remote Monitoring Tools

Various remote monitoring tools are available in the market, each offering unique features to cater to different needs. These tools typically include user-friendly interfaces accessible through web platforms or mobile applications. They provide a dashboard where users can visualize real-time data and historical trends and even receive alerts in case of system abnormalities.

Some advanced monitoring tools offer predictive analysis, using machine learning algorithms to forecast potential issues based on historical data and system parameters. This proactive approach empowers system owners to take preventive measures, minimizing downtime and maximizing energy production.

Diagnosing Common Solar System Issues and Failures

A fundamental prerequisite for effective troubleshooting is a comprehensive understanding of the different components that constitute an off-grid solar system. These components include solar panels, charge controllers, batteries, inverters, and the overall electrical wiring.

Solar panels are the heart of the system, responsible for converting sunlight into electricity. Charge controllers regulate the flow of energy from the panels to the batteries, preventing overcharging or deep discharging. Batteries store the generated energy for use during periods of low sunlight. Inverters convert the stored DC energy into AC, suitable for powering household appliances.

Identifying and Resolving Issues

Off-grid solar systems can face issues ranging from minor glitches to critical failures. A systematic approach is essential to diagnose and resolve these problems effectively.

1. **Energy Production Drop:** If the system's energy production decreases suddenly, it could indicate shading, dirt accumulation, or module malfunction. Visual inspection or using shade analysis tools can identify shading issues. Cleaning panels and replacing malfunctioning modules can restore energy output.

2. **Battery Problems:** Battery-related issues can lead to energy storage and supply problems. If batteries are not holding their charge or are overcharging, it might be due to faulty charge controllers or battery degradation. Voltage testing and assessing charge controller settings can help pinpoint the problem.

3. **Inverter Failures:** Inverters are crucial for converting energy into usable forms. A malfunctioning inverter can disrupt the entire system's performance. Monitoring inverter logs and performing diagnostic tests can help determine whether the issue lies with the inverter itself or other components.

4. **Wiring and Connection Faults:** Faulty wiring and poor connections can result in power loss or electrical hazards. Regularly inspecting and maintaining the wiring infrastructure can prevent such issues. Using thermal imaging cameras can identify overheating connections.

5. **Load Consumption Anomalies:** Unexpected load spikes or irregular consumption patterns might point to appliance malfunctions or wiring issues within the household. Ensuring that appliances are in good working condition and assessing load distribution can address such problems.

6. **Environmental Factors:** Harsh weather conditions like extreme temperatures or heavy snow can impact system performance. Designing systems to withstand such conditions and implementing protective measures can mitigate their effects.

Maintenance Best Practices to Prolong System Lifespan

Just like any mechanical system, off-grid solar systems require regular maintenance to ensure optimal performance and longevity. Regularly cleaning solar panels is paramount to remove dirt, debris, bird droppings, or any other obstructions that can hinder sunlight absorption. Cleaning can be done using water, a soft brush, or specialized cleaning solutions.

Apart from cleaning, routine visual inspections should be conducted to identify signs of wear and tear, corrosion, loose connections, or physical damage. Early detection of such issues can prevent them from escalating and causing significant system disruptions.

Batteries are a crucial element of off-grid solar systems, and their proper maintenance is essential to prolong their lifespan. This involves checking battery voltage levels, specifically gravity, and ensuring proper ventilation to prevent overheating. Regularly equalizing and topping up flooded lead-acid batteries with distilled water can extend their operational life.

Inverters and charge controllers should be inspected for any unusual noises, heat emissions, or error codes. Regular firmware updates, if applicable, can enhance their performance and compatibility with other system components.

While some maintenance tasks can be carried out by system owners, engaging professionals for comprehensive system inspections is recommended at least once a year. These professionals can conduct in-depth assessments, perform electrical tests, and make any necessary adjustments to optimize the system's efficiency.

Maintaining a detailed record of all maintenance activities, repairs, and component replacements is crucial. This documentation provides valuable insights into the system's history, helping technicians identify recurring issues and track the system's overall health.

BOOK 6

OFF-GRID SOLAR POWER FOR TINY HOMES

CHAPTER 1: SOLAR-POWERED HEATING AND COOLING SOLUTIONS

In an era where sustainability is no longer a buzzword but a necessity, harnessing renewable energy sources for our daily needs has become a paramount goal. Among these, solar power stands out as one of the most abundant and accessible resources. Solar energy offers a way to generate electricity and holds immense potential for addressing our heating and cooling requirements. This chapter delves into the fascinating realm of solar-powered heating and cooling solutions, exploring the various technologies and approaches that allow us to tap into the sun's energy for a more efficient and sustainable future.

Passive Solar Design: Harnessing Sunlight for Heating

The concept of passive solar design is rooted in the ingenious utilization of a building's architecture and materials to naturally collect, store, and distribute solar energy for heating purposes. This approach relies on optimizing a structure's orientation, layout, and features to passively capture and retain heat from the sun. By effectively using design elements such as windows, thermal mass, and insulation, passive solar design minimizes the need for mechanical heating systems, thereby reducing energy consumption and greenhouse gas emissions.

A fundamental principle of passive solar design is maximizing a building's south-facing exposure, where the majority of sunlight is available during the day. South-facing windows are strategically positioned to allow sunlight to enter the interior space. To prevent overheating in the summer months, roof overhangs or shading devices are incorporated to block the high-angle sun rays. During the colder months, the lower sun angle permits sunlight to penetrate deeper into the living spaces, providing a natural and comforting warmth.

Thermal mass, often in the form of dense materials like concrete or brick, is another crucial element in passive solar design. These materials have the ability to absorb and store heat, releasing it slowly over time. This helps regulate indoor temperatures by preventing rapid temperature fluctuations. An example of this can be observed in a sunlit room with a concrete floor – during the day, the floor absorbs heat, and during the night, it gradually releases the stored heat, maintaining a relatively stable temperature.

Furthermore, proper insulation is essential to prevent heat loss during colder periods. Well-insulated walls, roofs, and floors create a barrier that minimizes the transfer of heat from inside to outside, ensuring that the heat gained during the day remains within the building. This synergy of elements in passive solar design showcases how thoughtful architectural planning can significantly reduce the need for conventional heating systems.

Solar Space Heaters and Radiant Floor Heating

While passive solar design addresses the overall heating of a building, solar space heaters and radiant floor heating systems provide more targeted and controllable solutions. Solar space heaters are active solar systems that use sunlight to directly heat air or liquid, which is then circulated through a building to provide warmth. These systems consist of solar collectors – typically mounted on rooftops – that absorb solar radiation and convert it into heat energy. The collected heat is transferred to air or liquid, which is then distributed through the building's ventilation or radiant systems.

Radiant floor heating is a particularly comfortable and energy-efficient method of space heating. In this system, water is heated using solar energy and circulated through a network of pipes embedded in the floor. The heat radiates upward from the floor, warming the surrounding space evenly. This gentle and consistent heat distribution eliminates the discomfort of cold spots often experienced with traditional heating systems. Additionally, radiant floor heating operates at lower temperatures, making it compatible with solar heating and reducing energy consumption.

One of the advantages of solar space heaters and radiant floor heating is their modularity. These systems can be designed to suit various building sizes and layouts, making them adaptable for both residential and commercial applications. However, their efficiency depends on factors such as solar collector efficiency, insulation quality, and overall system design. Regular maintenance and proper system sizing are essential to ensure optimal performance.

Solar Air Conditioners and Ventilation Systems

As global temperatures rise, the demand for air conditioning has surged, contributing to higher electricity consumption. Solar air conditioners offer an eco-friendly alternative by utilizing the sun's energy to power cooling systems. These systems work on the principle of absorption refrigeration or desiccant cooling. In absorption refrigeration, solar heat is used to drive a chemical process that cools the air. Desiccant cooling involves using solar energy to remove moisture from the air, thereby reducing its temperature.

Absorption refrigeration systems use solar thermal collectors to heat a mixture of water and a refrigerant solution. As the solution evaporates due to the heat, it creates a cooling effect. The vaporized refrigerant is then condensed back to liquid form using a separate circuit cooled by outdoor air or water. This process of evaporation and condensation results in air conditioning without the need for traditional energy-intensive compressors.

On the other hand, desiccant cooling relies on moisture-absorbing materials to remove humidity from the air. Solar energy powers the regeneration of these materials, allowing them to continuously absorb moisture and cool the air. This method is particularly suitable for hot and humid climates where traditional air conditioning struggles to maintain comfort efficiently.

Solar-powered ventilation systems also play a crucial role in maintaining indoor air quality and temperature. These systems use solar energy to power fans that draw in fresh outdoor air and expel stale indoor air. Heat recovery ventilation (HRV) and energy recovery ventilation (ERV) systems further enhance efficiency by exchanging heat between the incoming and outgoing air streams. This ensures that a significant portion of the energy used to condition the air is retained, reducing the overall energy demand.

CHAPTER 2: SOLAR WATER HEATING SYSTEMS FOR TINY HOMES

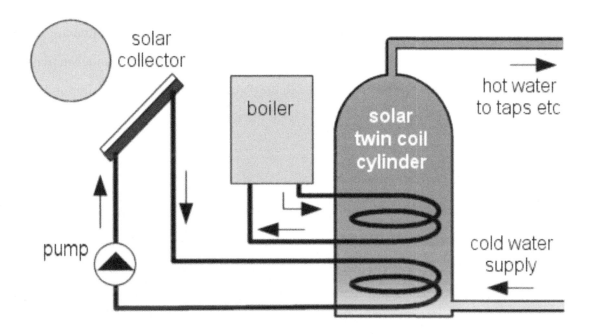

In the pursuit of sustainable living and reduced energy consumption, tiny homes have gained immense popularity as an innovative solution. With their compact size and eco-friendly design, these homes are environmentally conscious and economical to maintain. One crucial aspect of sustainable living is the utilization of renewable energy sources, and solar water heating systems have emerged as a game-changer in this regard. This chapter delves into the world of solar water heating systems specifically tailored for tiny homes, exploring various types of solar water heaters, DIY construction, and installation methods, and the role of solar water pumps in off-grid systems.

Types of Solar Water Heaters: Batch, Flat-Plate, and Evacuated Tube

Solar water heaters are ingenious devices that harness the power of the sun to provide a cost-effective and eco-friendly solution for heating water. Three primary types of solar water heaters are commonly employed, each with its own distinct design and advantages.

Batch Solar Water Heater

The batch solar water heater, often referred to as the "Integrated Collector and Storage" system, is a simple and effective design. It consists of a black storage tank, usually positioned inside an insulated box with a glass cover. The black color of the tank absorbs solar energy and heats the water within. This design is suitable for regions with moderate climates, as it may be susceptible to freezing in colder climates due to its exposed plumbing.

Flat-Plate Solar Water Heater

Flat-plate solar water heaters are more complex systems that consist of an insulated box containing a dark-colored absorber plate covered by glass or plastic. The absorber plate heats up as sunlight strikes it and then transfers the heat to the water that flows through pipes or tubes connected to the plate. These heaters are versatile and can be used in various climates, as they can incorporate antifreeze to prevent freezing.

Evacuated Tube Solar Water Heater

Evacuated tube solar water heaters are the most efficient and versatile option. These systems consist of rows of glass tubes, each containing a metal absorber tube and a vacuum. The vacuum acts as insulation, reducing heat loss and making these heaters highly efficient even in cold climates. The absorber tubes heat up rapidly, transferring the heat to a fluid within the tubes, which then transfers the heat to the water. This design ensures excellent performance in all weather conditions.

Choosing the right type of solar water heater for a tiny home depends on factors such as climate, available space, budget, and aesthetic preferences. While batch heaters are straightforward and cost-effective, flat-plate and evacuated tube heaters offer greater efficiency and adaptability.

DIY Solar Water Heater Construction and Installation

The concept of "do-it-yourself" (DIY) solar water heaters has gained traction as people seek to reduce costs and actively participate in sustainable practices. Constructing and installing a DIY solar water heater for a tiny home requires careful planning, some technical skills, and the right materials. Here's a step-by-step guide to creating a basic batch solar water heater:

1. **Step 1: Gather Materials and Tools.** To build a batch solar water heater, you'll need a black storage tank, insulation material, an insulated box, a glass or plastic cover, pipes or tubing, and plumbing fittings. Tools like a saw, drill, screws, and caulking gun will also be necessary.

2. **Step 2: Build the Insulated Box** Construct an insulated box that will house the storage tank. The box should have enough space for proper insulation around the tank. Ensure one side of the box can be opened and closed, allowing access to the tank.

3. **Step 3: Install the Storage Tank.** Place the black storage tank inside the insulated box. The tank should be positioned to maximize sunlight exposure. Attach pipes or tubing to the tank's inlet and outlet.

4. **Step 4: Add Insulation.** Fill the space around the storage tank with insulation material. This prevents heat loss and maintains water temperature.

5. **Step 5: Add the Cover.** Attach the glass or plastic cover to the open side of the insulated box. This cover allows sunlight to enter while trapping heat inside.

6. **Step 6: Connect Plumbing** Connect the inlet and outlet pipes or tubing to the water supply and the point of use within the tiny home. Ensure proper sealing with plumbing fittings and caulking.

7. **Step 7: Test the System** Fill the tank with water, and the solar water heater is ready to use. Monitor the temperature over a few days and make adjustments as needed.

Creating a DIY solar water heater demands careful attention to detail and safety. While the batch system outlined above is relatively simple, flat-plate and evacuated tube systems require more intricate construction. It's essential to thoroughly research and understand the chosen design before embarking on the DIY journey. Moreover, consulting professionals or resources specialized in solar heater construction can be immensely helpful.

Solar Water Pumps and Circulation in Off-Grid Systems

For tiny homes located in off-grid settings, achieving efficient circulation of water within the solar water heating system is crucial. Solar water pumps play a pivotal role in maintaining a steady flow of water, ensuring effective heat transfer and distribution. Here's an exploration of their significance and how they function:

Solar water pumps are electric pumps powered by solar panels. They are responsible for circulating water through the solar collectors (absorber plates or tubes) and then into the storage tank or point of use. This circulation prevents stagnation and ensures that heated water is appropriately distributed.

Importance of Solar Water Pumps:

1. **Efficient Heat Transfer:** The constant circulation of water prevents hot water from becoming trapped in the collectors, where it could lose heat to the surroundings. This efficiency translates to hotter water available for use.

2. **Prevention of Freezing:** In cold climates, stagnant water within the collectors can freeze and damage the system. Even when the sun is not shining, continuous circulation mitigates this risk.

3. **Balanced System:** Solar water pumps help maintain consistent temperatures throughout the system. This is particularly essential in larger installations or systems with multiple collectors.

How Solar Water Pumps Work:

1. **Photovoltaic (PV) Panels:** Solar water pumps are connected to PV panels that convert sunlight into electricity. This electricity powers the pump, eliminating the need for an external power source.

2. **Controller:** A controller manages the pump's operation, regulating its speed or turning it on and off based on temperature differentials or user-defined settings.

3. **Sensors:** Temperature sensors placed at strategic points, such as the collectors and the storage tank, send signals to the controller. The controller uses this information to determine when to activate the pump.

4. **Circulation:** The pump starts when the temperature differential between the collectors and the tank reaches a certain threshold. It pushes water through the collectors, where it absorbs heat, and then directs it to the storage tank or point of use.

Solar water pumps can vary in size and capacity based on the system's complexity and the volume of water to be circulated. In off-grid tiny homes, where conventional power sources might be limited, solar water pumps offer an elegant solution that aligns perfectly with the overall sustainable approach of the dwelling.

Solar water heating systems have found a perfect match in the world of tiny homes. These compact dwellings, designed with efficiency and sustainability in mind, benefit greatly from the innovative technology of solar water heaters. Whether through the simplicity of batch heaters, the versatility of flat-plate systems, or the efficiency of evacuated tube heaters, there's a solar solution for every tiny home's unique needs.

For those eager to take an active role in creating their energy solutions, DIY solar water heater construction opens up opportunities for customization and cost savings. However, the complexity of such projects demands careful planning, research, and potentially expert guidance.

In off-grid scenarios, solar water pumps step in as the silent heroes, ensuring proper circulation and efficient heat transfer. These pumps, powered by sunlight, exemplify the harmony between technology and nature that defines the essence of sustainable living.

As the world continues to explore innovative ways to reduce carbon footprints and embrace renewable energy, the fusion of solar water heating systems and tiny homes stands as a testament to human creativity and the pursuit of a greener future.

CHAPTER 3: SOLAR-POWERED APPLIANCES AND GADGETS

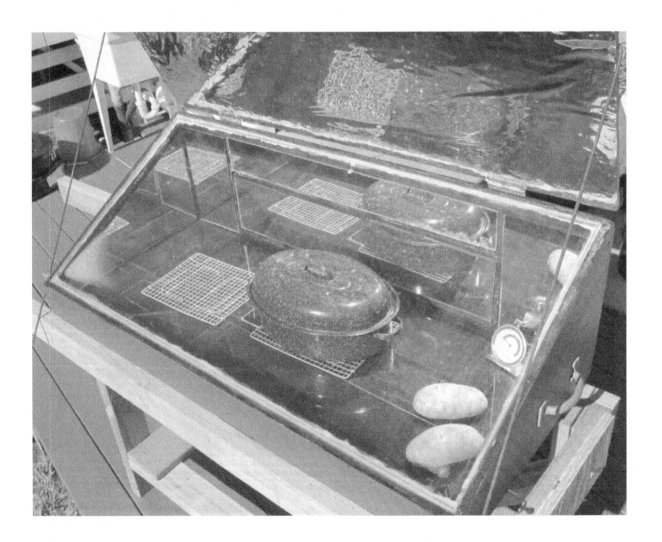

In an era defined by advancing technology and growing environmental concerns, the harnessing of solar power has emerged as a transformative solution. Solar energy, derived from the sun's radiant light and heat, has found applications across various sectors, from electricity generation to heating and cooling systems. One of the most intriguing and impactful manifestations of solar power's potential lies in the realm of everyday appliances and gadgets. This chapter delves into the realm of solar-powered appliances and gadgets, exploring their significance, functionality, and the ways in which they contribute to sustainable living.

Solar Ovens and Cooking Appliances

Cooking is an essential aspect of human life, and with the integration of solar power, this mundane yet vital activity has been elevated to a sustainable and innovative level. Solar ovens and cooking appliances represent a remarkable fusion of ancient practices and modern technology. The concept behind solar cooking is elegantly simple: harness the sun's energy to generate heat for cooking without relying on fossil fuels or electricity.

Solar ovens come in various designs, but their underlying principle remains consistent. A solar oven typically consists of a reflective surface, which concentrates sunlight onto a cooking chamber, and insulation to trap the generated heat. This concentration of sunlight leads to temperatures sufficient for cooking a variety of dishes, showcasing the adaptability of this approach. From simple designs that individuals can construct themselves using cardboard and aluminum foil to more advanced commercially available models, solar ovens offer an eco-friendly alternative to conventional cooking methods.

The benefits of solar ovens extend beyond sustainability. They are particularly advantageous in regions with unreliable access to electricity or where traditional fuel sources are scarce. Developing countries, for instance, have embraced solar ovens as a means to address energy poverty and reduce indoor air pollution caused by traditional cooking methods. Solar cooking also has health benefits, as it eliminates the harmful fumes emitted by burning wood or charcoal.

Solar-Powered Refrigerators and Freezers

Refrigeration is another area where solar power has made substantial inroads. Refrigerators and freezers are indispensable for preserving food and medicines, but their operation traditionally relies on electricity, which may not always be accessible, especially in remote or off-grid locations. Solar-powered refrigeration presents an innovative solution to this challenge.

Solar refrigerators and freezers operate on the same fundamental principle as traditional units, but they integrate photovoltaic (PV) panels to capture solar energy and convert it into electricity to power the cooling system. These appliances are equipped with energy storage solutions, such as batteries, to ensure continuous operation even during periods of low sunlight.

The advantages of solar-powered refrigeration are manifold. Beyond addressing energy accessibility issues, these appliances contribute to reducing carbon emissions by minimizing reliance on fossil fuel-based electricity. In regions prone to power outages, solar refrigeration ensures the preservation of perishable goods, reducing food waste and economic losses. Furthermore, solar-powered refrigerators find applications in the medical field, enabling the safe storage of vaccines and medicines in remote clinics or disaster-stricken areas.

Solar-Powered Electronics and Charging Solutions

The ubiquity of electronic devices in modern society and the need to keep them charged presents both a challenge and an opportunity. Solar-powered electronics and charging solutions have emerged as a response to this need, revolutionizing the way we interact with technology.

Portable solar chargers and solar-powered battery banks have gained popularity due to their convenience and environmental benefits. These devices consist of solar panels that convert sunlight into electricity, which is stored in integrated batteries. The stored energy can then be used to charge smartphones, tablets, laptops, and other gadgets. This innovation is particularly advantageous for outdoor enthusiasts, travelers, and individuals living in regions with unreliable electricity grids.

Beyond individual charging solutions, solar power has also influenced the design of public spaces and infrastructure. Solar-powered charging stations have been installed in parks, airports, and urban centers, offering the public access to clean energy for recharging their devices. These stations provide a valuable service and raise awareness about the potential of renewable energy sources.

The integration of solar power into everyday appliances and gadgets marks a pivotal juncture in sustainable living and technological innovation. Solar ovens and cooking appliances reimagine the way we prepare meals, offering a clean and efficient alternative to traditional cooking methods. Solar-powered refrigerators and freezers address energy accessibility and conservation challenges, ensuring the reliable preservation of perishable goods. Solar-powered electronics and charging solutions cater to the growing demand for portable energy, transforming the way we power our devices while reducing our carbon footprint.

These developments underscore solar energy's transformative potential beyond its traditional electricity generation applications. As technology continues to evolve, the efficiency and affordability of solar-powered appliances are likely to increase, making them even more accessible to a wider population. The convergence of sustainability and convenience within the realm of solar-powered appliances and gadgets paves the way for a future where eco-friendly solutions seamlessly integrate into our daily lives, reshaping our relationship with both technology and the environment.

CHAPTER 4: OFF-GRID LIGHTING: LED AND OTHER SOLUTIONS

In a world increasingly defined by the quest for sustainability and energy efficiency, the realm of lighting technology has witnessed remarkable innovations that have transformed our physical environments and attitudes toward energy consumption. One prominent aspect of this evolution is the advent of off-grid lighting solutions, a topic that encapsulates the fusion of energy-efficient LED lighting and the ingenious harnessing of solar power. This chapter delves into the multifaceted landscape of off-grid lighting, particularly focusing on LED lighting for tiny spaces, solar lighting systems for indoor and outdoor applications, and the captivating realm of solar-powered pathway lights and security lighting. By exploring these dimensions, we unravel these illuminating innovations' profound impacts, technological intricacies, and real-world applications.

Energy-Efficient LED Lighting for Tiny Spaces

The evolution of lighting technology took a monumental leap with the introduction of Light Emitting Diodes (LEDs). These tiny yet immensely powerful semiconductor devices have revolutionized the way we illuminate spaces, offering unparalleled energy efficiency, durability, and versatility. Their efficiency is derived from the conversion of electrical energy directly into light, circumventing the wasteful heat production characteristic of traditional incandescent bulbs.

In tiny spaces, such as compact apartments, cozy cabins, or recreational vehicles, LED lighting emerges as the perfect companion. The compactness of LEDs allows for creative integration into limited spaces without compromising on illumination quality. Their low power consumption is particularly advantageous when energy sources are scarce, making them ideal for off-grid lighting setups.

Beyond their technical attributes, LED lighting also offers an aesthetic advantage. The controllable nature of LED color and intensity opens the doors to personalized lighting schemes that can adapt to various moods and activities. This feature is particularly beneficial in tiny spaces where each corner serves multiple functions throughout the day.

However, the successful integration of LED lighting in tiny spaces requires thoughtful design and placement. A balance between fixture placement, beam angle, and diffusion is crucial to avoid issues like glare and uneven illumination. Additionally, the consideration of color temperature plays a pivotal role, as it influences the ambiance of the space.

Solar Lighting Systems: Indoor and Outdoor Applications

The quest for sustainability and reduced carbon footprint has propelled the solar energy industry into the limelight. Solar panels initially harnessed for electricity generation, have found an intriguing niche in the realm of lighting systems. The marriage of solar panels and efficient LED lighting has given birth to solar lighting systems with diverse indoor and outdoor applications.

Solar lighting systems offer an innovative solution for spaces with limited access to traditional power sources in indoor settings. Think of a remote cabin tucked away in the woods or an emergency shelter during a power outage. These systems comprise solar panels that harvest sunlight and store it in batteries for nighttime use. Coupled with energy-efficient LED bulbs, these setups ensure a sustainable and reliable lighting source.

Solar lighting systems have revolutionized landscape and architectural illumination on the outdoor front. Solar-powered garden lights, for instance, rely on photovoltaic cells to accumulate solar energy during the day, subsequently casting an enchanting glow over pathways and plant beds at night. The elimination of wires and the absence of dependence on the grid render these solutions aesthetically pleasing and remarkably easy to install.

Solar-Powered Pathway Lights and Security Lighting

The interplay of solar power and LED lighting becomes especially enchanting when it comes to solar-powered pathway lights and security lighting. Picture a moonlit garden path illuminated by soft, ethereal lights that draw inspiration from the sun. Solar-powered pathway lights epitomize the perfect harmony between form and function.

These lights typically consist of small solar panels that discreetly blend into the surrounding landscape. During the day, they soak up sunlight, charging internal batteries. As dusk descends, built-in sensors trigger the LED bulbs to illuminate the pathway, creating a captivating visual effect while ensuring safety.

Equally compelling is the application of solar power in security lighting. Traditional security lighting often relies on power-hungry fixtures that can strain electricity resources. Solar-powered security lights, on the other hand, offer an eco-friendly alternative that doesn't compromise on effectiveness. Whether illuminating a dimly lit alleyway or accentuating the entrance of a home, these lights harness sunlight, converting it into a powerful deterrent against intruders.

However, like any technological innovation, there are challenges to overcome. One primary concern with solar-powered lighting is its dependence on sunlight. In regions with inconsistent sunlight or during prolonged cloudy periods, the efficiency of these systems can be compromised. Additionally, the initial investment required for solar panels and high-quality LED fixtures can be a deterrent for some individuals or communities despite the long-term savings.

In a world characterized by environmental consciousness and the pursuit of innovative solutions, off-grid lighting stands as a testament to human ingenuity. The amalgamation of energy-efficient LED technology with the boundless power of the sun has birthed a new era of lighting solutions that are both efficient and enchanting. Solar-powered lighting has carved a niche that seamlessly merges sustainability, functionality, and aesthetics from the cozy confines of tiny spaces to the expansive landscapes of gardens and pathways.

As technology continues to advance and our understanding of energy systems deepens, the trajectory of off-grid lighting appears promising. Researchers and innovators are continually refining solar panels' efficiency, battery storage capabilities, and LED lighting performance. This relentless pursuit ensures that the realm of off-grid lighting will only become more accessible and impactful, further cementing its place in the grand narrative of sustainable living.

BOOK 7

OFF-GRID SOLAR
POWER FOR CABINS

CHAPTER 1: DESIGNING SOLAR-POWERED CABIN RETREATS

Designing Solar-Powered Cabin Retreats is a pursuit that harmonizes the natural world with modern comfort and sustainability. The concept of off-grid cabin living has grown as a response to the hectic pace of contemporary life, offering an opportunity to disconnect and embrace a more ecologically sensitive lifestyle. This chapter delves into the multifaceted considerations and practices that go into creating solar-powered cabins that seamlessly blend with the environment while providing a comfortable and aesthetically pleasing retreat.

Off-Grid Cabin Design Considerations and Aesthetics

In the process of designing an off-grid solar-powered cabin, one of the most crucial elements is site selection and integration. Choosing the right location and positioning the cabin thoughtfully can significantly impact both energy efficiency and the overall experience. A comprehensive understanding of the site's topography, solar exposure, prevailing winds, and vegetation is fundamental. This knowledge informs architects and designers on how to best position the cabin for optimal solar energy utilization while minimizing its impact on the surrounding landscape.

Aesthetic considerations play an equally important role in cabin design. The modern cabin aesthetic revolves around the idea of "rustic modernism," blending traditional cabin warmth with contemporary design elements. The use of materials like reclaimed wood, stone, and metal helps create a sense of coziness while maintaining a sleek and minimalistic appeal. Large windows and open floor plans are not only aesthetic choices but also practical ones, as they allow for a seamless connection with nature and provide a feeling of spaciousness within the confined cabin space.

Space optimization is a critical aspect of cabin design due to the limited square footage. Multipurpose furniture, built-in storage solutions, and open shelving are clever strategies to maximize usable space while maintaining an uncluttered environment. The design process involves finding a delicate balance between functionality and aesthetics to ensure that the cabin remains both comfortable and visually appealing.

When it comes to materials, sustainable choices are of paramount importance. Opting for environmentally friendly materials, such as low VOC paints and bamboo flooring, aligns with the principles of sustainability. Integrating recycled or upcycled materials not only reduces the cabin's carbon footprint but also imparts a unique character that reflects the cabin's commitment to eco-conscious living.

Remote Cabin Power Supply: Determining Energy Needs

Determining the energy requirements of a remote cabin is the first step toward establishing a reliable and efficient solar power system. Conducting an energy audit involves assessing the power needs of various components within the cabin, including lighting, heating, cooling, appliances, and electronics. This comprehensive evaluation serves as the foundation for designing a solar power system tailored to the specific energy demands of the cabin.

The feasibility of solar power for a remote cabin hinges on several factors, including geographical location, orientation, and local climate. Technological advancements in solar panel efficiency allow architects and designers to estimate energy production with a high degree of accuracy. Tools such as solar irradiance maps aid in calculating the amount of solar energy available throughout the year, enabling more precise system design and performance predictions.

One of the challenges of solar power generation is its intermittency. To address this, battery storage systems are essential for storing excess energy generated during sunny periods for use during cloudy days or at night. Selecting the right type and capacity of batteries is crucial to ensuring optimal system performance and longevity.

While solar power systems aim to meet most of a cabin's energy needs, having a backup generator is a prudent measure, particularly during extended periods of low sunlight. Backup generators, often powered by propane or biodiesel, provide supplementary power when solar generation is insufficient, ensuring uninterrupted energy supply for essential functions.

Balancing Comfort and Sustainability in Cabin Living

Achieving comfort in a solar-powered cabin extends beyond aesthetics. Passive solar design is a cornerstone of energy-efficient cabin living. It leverages the movement of the sun to naturally heat and cool the cabin, reducing the reliance on mechanical systems. Thoughtful placement of windows, incorporation of thermal mass, and proper insulation work in synergy to maintain comfortable indoor temperatures throughout the year.

Cabin living necessitates mindful choices in terms of appliances. Opting for energy-efficient appliances not only reduces power consumption but also aligns with the cabin's overarching commitment to sustainability. From LED lighting to propane-powered refrigerators and low-energy heating and cooling systems, these choices contribute to the efficient utilization of energy resources.

Incorporating intelligent energy management systems further enhances both comfort and sustainability. Smart thermostats, lighting controls, and energy monitoring systems empower cabin occupants to regulate energy consumption efficiently. The ability to control these systems remotely through smartphones ensures optimal energy management even when away from the cabin.

Cabin living often encourages a minimalist lifestyle, a philosophy that resonates with the sustainable ethos of off-grid living. With limited space, inhabitants are prompted to declutter and prioritize essential belongings, leading to reduced consumption patterns and a deeper connection with the surrounding natural environment.

Sustainability isn't confined to the cabin's interior. Thoughtful landscaping practices, such as the use of native plants and rain gardens, minimize water usage and seamlessly integrate the cabin into its surroundings. These landscaping choices contribute to the overall ecological balance of the cabin retreat.

CHAPTER 2: CABIN OFF-GRID PLUMBING SOLUTIONS

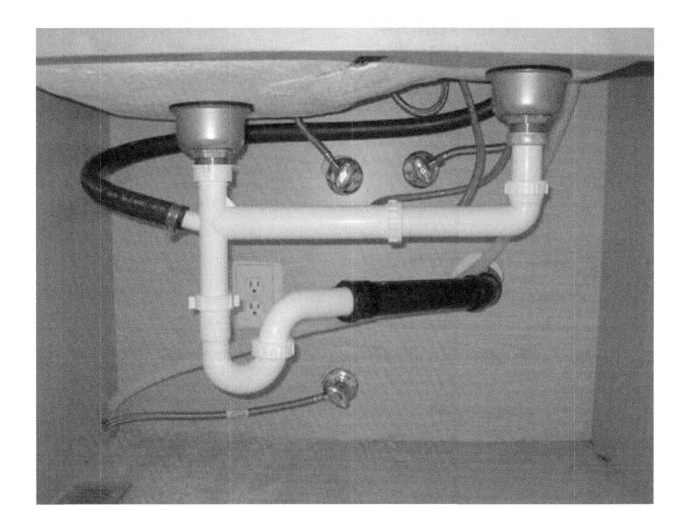

In the tranquil embrace of nature, far from the bustling urban landscape, cabins offer a haven of solitude and serenity. Yet, the allure of an off-grid cabin comes with its unique set of challenges, particularly in terms of plumbing solutions. Cabin owners seeking a harmonious existence with nature while enjoying modern comforts often find themselves exploring innovative plumbing methods that minimize environmental impact and maximize self-sufficiency. This chapter delves into the realm of cabin-off-grid plumbing solutions, where the marriage of technology and eco-consciousness paves the way for sustainable living.

Rainwater Harvesting and Greywater Recycling

Rainwater harvesting stands as a testament to the ingenious ways humans can harmonize with the natural world. In the context of off-grid cabins, this practice involves collecting rainwater from rooftops and other surfaces to be utilized for various non-potable purposes. By doing so, cabin owners can significantly reduce their reliance on traditional water sources, all while fostering a deeper connection with the environment.

One of the key components of a rainwater harvesting system is the collection infrastructure. Cabin roofs are designed to efficiently channel rainwater into gutters and downspouts, which in turn direct the water into storage tanks. These tanks, often constructed from durable materials such as polyethylene or concrete, act as reservoirs for rainwater. To prevent contamination, a "first flush" mechanism is employed, diverting the initial flow of rainwater, which might carry debris or pollutants away from the storage tanks.

The harvested rainwater finds a multitude of applications within an off-grid cabin. It can be employed for irrigation purposes, nurturing gardens, and cultivating homegrown produce. Additionally, rainwater serves as an excellent source for flushing toilets, reducing the burden on septic systems. Some adventurous cabin owners even embrace the idea of installing outdoor showers that utilize rainwater, allowing for a truly immersive bathing experience.

Greywater, the relatively clean wastewater generated from activities such as washing dishes, doing laundry, and bathing, holds untapped potential in the realm of off-grid cabin plumbing. Greywater recycling involves the treatment and reuse of this water for purposes other than drinking, such as irrigation or flushing toilets. By adopting such a system, cabin owners can dramatically reduce water waste and further their commitment to sustainable living.

A common greywater recycling setup includes a filtration system that purifies the wastewater, rendering it suitable for non-potable applications. Filters, such as sand and gravel beds, remove impurities and particulates, while biological processes break down organic matter. This treated greywater is then stored in separate tanks and can be used for watering plants or replenishing toilet reservoirs.

The adoption of greywater recycling not only conserves water but also fosters a heightened awareness of consumption patterns. Cabin dwellers become attuned to their usage habits, exploring ways to minimize water wastage and maximize efficiency. In essence, greywater recycling transforms a simple cabin into a classroom of conscientious consumption.

Solar-Powered Water Pumps and Filtration Systems

In the realm of off-grid living, self-sufficiency is often epitomized by the utilization of renewable energy sources. Solar power, with its abundance and eco-friendliness, takes center stage. When applied to water systems, it yields solar-powered water pumps that provide a lifeline for cabins far removed from traditional power grids.

Solar water pumps function by converting sunlight into electrical energy, which powers the pumping mechanism. This energy is stored in batteries, ensuring a continuous water supply even during cloudy days or at night. These pumps can draw water from wells, springs, or other natural sources, delivering it to storage tanks for distribution throughout the cabin.

The marriage of solar power and water pumps liberates cabin owners from the constraints of conventional electricity. It offers a seamless blend of technology and nature, where the sun's rays fuel the rhythms of daily life. The environmental benefits are substantial, as solar-powered pumps produce no greenhouse gas emissions and have minimal maintenance requirements.

Access to clean, potable water is an undeniable necessity. Off-grid cabins, often nestled in remote locales, might lack access to municipal water treatment facilities. In such scenarios, advanced filtration systems emerge as the safeguard against waterborne contaminants, ensuring a healthy and secure water supply.

Filtration systems designed for off-grid cabins are characterized by their efficiency and reliability. These systems typically consist of multiple stages, each designed to eliminate specific impurities. Sediment filters remove larger particles, while activated carbon filters adsorb organic compounds and chlorine. More advanced systems incorporate reverse osmosis, a process that leverages semipermeable membranes to eliminate microscopic contaminants, ensuring the water's purity.

The integration of filtration systems into off-grid plumbing exemplifies the synergy between human innovation and environmental stewardship. Cabin owners can savor the assurance of safe drinking water while simultaneously lessening the burden on surrounding ecosystems. This harmony extends the cabin's footprint beyond its physical boundaries, fostering a ripple effect of sustainable practices.

Composting Toilets and Waste Management

The conventional concept of waste takes on a new dimension in the world of off-grid cabins, where the harmonious cycle of nature extends to the realm of sanitation. Composting toilets are a striking embodiment of this philosophy, transforming human waste into a valuable resource and eliminating the need for complex sewage systems.

Composting toilets operate on a simple principle: waste is mixed with organic materials like sawdust or coconut coir, creating an environment conducive to decomposition. This process, facilitated by microbes and bacteria, breaks down the waste into nutrient-rich compost over time. This compost can then be safely used as fertilizer for non-edible plants, closing the loop in a remarkably sustainable manner.

Beyond the ecological advantages, composting toilets offer practical benefits for off-grid cabin owners. They require minimal water, addressing the water scarcity challenge faced by many cabins. Moreover, the absence of sewage infrastructure simplifies construction and maintenance, making these toilets a practical and efficient solution for remote living.

Waste management in an off-grid cabin encompasses more than just human waste. It extends to a holistic approach to waste reduction, recycling, and proper disposal. Cabin dwellers, driven by a reverence for nature, embrace the principles of "leave no trace" and "zero waste," ensuring their environmental impact remains minimal.

Recycling systems are integral to this waste management philosophy. Separate bins for different types of recyclables empower cabin owners to minimize landfill-bound waste. Composting, beyond toilet waste, extends to kitchen scraps and organic materials, nourishing the soil rather than burdening it.

CHAPTER 3: OFF-GRID REFRIGERATION AND FOOD STORAGE

In a world that is becoming increasingly reliant on technology and energy, the concept of off-grid living has gained significant traction. This shift towards self-sustainability and reduced reliance on traditional power sources has led to a reexamination of various aspects of daily life, including food storage and refrigeration. As we delve into Chapter 3 of our exploration of off-grid living, we will dissect the nuances of off-grid refrigeration and innovative food storage techniques, from the age-old wisdom of root cellars to the modern marvels of solar food dehydrators.

Propane vs. Solar-Powered Refrigeration: Weighing the Options

The heart of any discussion on off-grid refrigeration revolves around the choice of power source. Two primary contenders in this arena are propane and solar power. Each comes with its unique set of advantages and challenges, catering to different needs and preferences of off-grid enthusiasts.

Propane-powered refrigeration has long been a staple for those seeking off-grid cooling solutions. Propane, a flammable hydrocarbon gas, is known for its efficiency and portability. Propane refrigerators operate by using a heat-absorption process that doesn't rely on traditional electric compressors. Instead, they utilize a combination of ammonia, hydrogen gas, and water. When heat is applied, the ammonia gas condenses into a liquid, absorbing heat from the refrigerator's interior and thereby cooling it.

One of the primary benefits of propane refrigeration is its reliability. Unlike solar power, which can be intermittent depending on weather conditions, propane-powered systems ensure constant cooling as long as there is a supply of propane gas. This is especially crucial in regions with unpredictable weather patterns or limited sunlight.

However, propane refrigeration comes with its share of drawbacks. Firstly, propane is a non-renewable fossil fuel, and its availability might be a concern in remote areas. Additionally, there are safety considerations due to the flammability of propane. Proper ventilation and monitoring systems are necessary to prevent potential hazards. Furthermore, the maintenance and repair of propane refrigerators can be complex, requiring specialized knowledge.

Solar-powered refrigeration represents a leap into the future of off-grid living. With advancements in solar panel technology, harnessing the sun's energy has become a viable option for powering essential appliances, including refrigerators. Solar-powered refrigerators operate similarly to their electric counterparts but rely on photovoltaic panels to convert sunlight into electricity.

The environmental benefits of solar power are undeniable. It is a clean, renewable energy source that reduces reliance on fossil fuels and decreases the carbon footprint. Solar-powered refrigeration systems require minimal maintenance, with no need for fuel replenishment. They also provide consistent cooling as long as there's sunlight available.

However, solar-powered refrigeration does have its limitations. It heavily relies on the availability of sunlight, making it less dependable in cloudy or rainy conditions, especially during winter months or in regions with reduced sunlight hours. The initial setup costs can be relatively high, including the price of solar panels, batteries, and inverters. Proper system design is crucial to ensure adequate energy storage for nighttime usage or cloudy days.

Root Cellars and Passive Food Storage Techniques: Nurturing Nature's Wisdom

Long before the advent of modern refrigeration, our ancestors devised ingenious methods for preserving food without electricity. Root cellars and passive food storage techniques stand as timeless examples of harnessing nature's resources to create effective off-grid food storage solutions.

Root cellars are subterranean structures used to store a variety of fruits, vegetables, and even canned goods. These cellars take advantage of the natural insulating properties of the earth, creating a cool and humid environment that helps extend the shelf life of perishable items.

The concept is simple yet effective. A well-built root cellar is typically located underground or partially buried, allowing it to tap into the stable temperatures of the earth. The surrounding soil acts as a natural insulator, keeping the interior of the cellar cool in the summer and preventing extreme freezing in the winter. Additionally, the humidity levels inside a root cellar help maintain the freshness of stored produce.

Constructing a root cellar requires careful consideration of factors such as location, ventilation, and moisture control. Proper shelving and organization are also essential to prevent spoilage and maximize storage capacity. While root cellars are effective, they do have limitations. They are best suited for cooler climates and might not be as effective in regions with consistently high temperatures.

Beyond root cellars, there are a plethora of passive food storage techniques that have stood the test of time. Drying, salting, fermenting, and pickling are just a few examples of age-old methods that preserve food without the need for electricity.

Drying, in particular, is a technique that has been used for centuries. By removing the moisture from food, microorganisms that cause spoilage and decay are inhibited. Fruits, vegetables, meats, and herbs can all be effectively dried using the sun's heat or gentle air circulation. The result is lightweight, shelf-stable products that retain much of their nutritional value.

Fermentation and pickling are other ingenious methods that not only preserve food but also enhance its flavor. Microorganisms responsible for the fermentation process create an environment that inhibits harmful bacteria while imparting unique tastes to the food. Sauerkraut, kimchi, and various types of pickles are classic examples of fermented and pickled foods.

Solar Food Dehydrators and Preservation Methods: Modern Twists on Timeless Techniques

As we embrace modern technology, innovative off-grid solutions have emerged that build upon traditional food preservation techniques. Solar food dehydrators exemplify the synergy of age-old wisdom and contemporary engineering.

Solar food dehydrators are designed to harness the sun's energy for the purpose of drying food. This process removes the moisture content, preventing the growth of microorganisms and preserving the food's nutritional value. Unlike electric dehydrators, their solar counterparts use renewable energy and have a minimal environmental impact.

The basic design of a solar food dehydrator includes trays for placing sliced or chopped food, a transparent cover to capture sunlight, and vents to facilitate airflow. As the sun's heat enters the dehydrator, it causes the moisture in the food to evaporate. The warm, moist air then exits the dehydrator through the vents, leaving behind perfectly preserved and lightweight edibles.

One of the significant advantages of solar food dehydrators is their simplicity. They require no electricity, making them an ideal solution for off-grid settings. Additionally, they can be constructed using readily available materials, often as DIY projects. Solar food dehydrators are particularly valuable for preserving seasonal abundance or creating on-the-go snacks for outdoor adventures.

While traditional preservation techniques continue to thrive, modern adaptations have brought convenience and efficiency to off-grid food storage. Canning, for instance, has evolved from the labor-intensive processes of the past to streamlined methods that retain the nutritional quality of the food.

Pressure canning and water bath canning are two widely used methods for preserving foods like fruits, vegetables, and meats. These techniques involve sealing food in airtight containers, preventing the growth of spoilage organisms. Pressure canning is especially useful for low-acid foods, while water bath canning is suitable for high-acid foods.

Vacuum sealing is another contemporary preservation method that removes air from packaging, significantly reducing the risk of spoilage and freezer burn. This technique is particularly popular for meats, as it maintains the flavor and quality of the products.

CHAPTER 4: SECURING AND PROTECTING YOUR CABIN SOLAR SYSTEM

In this pivotal chapter, we delve into the critical aspects of securing and safeguarding your cabin solar system. Your off-grid solar setup is a valuable investment, both in terms of energy production and your overall cabin experience. Ensuring its security and protection not only guarantees the longevity of your system but also contributes to the safety and functionality of your remote abode. This chapter will guide you through the multifaceted process of fortifying your cabin solar system, encompassing cabin security systems and surveillance, lightning and surge protection for solar arrays, and the indispensable practice of winterizing your off-grid cabin and solar equipment.

Cabin Security Systems and Surveillance

When you embark on the journey of living off-grid, the tranquility of remote living comes hand in hand with the responsibility of ensuring your cabin's security. Cabin security systems and surveillance mechanisms play a pivotal role in safeguarding both your property and the cabin solar system itself. Given the isolation of off-grid living, investing in robust security measures becomes paramount.

Security System Integration

The integration of a comprehensive security system is the first step towards fortifying your cabin. Modern security systems encompass a range of technologies, including motion sensors, door and window alarms, security cameras, and even smart home automation. These systems can be configured to notify you remotely of any potential breaches, enabling you to take immediate action.

Surveillance Cameras

Deploying surveillance cameras around your cabin and within proximity to your solar setup can significantly deter potential intruders. High-resolution cameras equipped with night vision capabilities offer round-the-clock monitoring. When strategically placed, these cameras not only provide visual evidence in case of an incident but also aid in the remote assessment of your cabin's security.

Remote Monitoring

With advancements in technology, remote monitoring of your cabin and solar system has become remarkably accessible. Real-time feeds from surveillance cameras can be accessed through your smartphone or computer. This feature is invaluable, especially when you're away from the cabin for extended periods. It ensures that you remain connected and informed about your property's security status at all times.

Alarm Systems

Alarm systems are your first line of defense against unauthorized access. Motion sensors strategically positioned around the cabin's perimeter and near the solar array can trigger audible alarms when any suspicious movement is detected. These alarms not only alert you but also act as a potent deterrent, discouraging intruders from proceeding further.

Lightning and Surge Protection for Solar Arrays

An often underestimated threat to your cabin's solar system is the potential for lightning strikes and power surges. Lightning poses a substantial risk, considering the cabin's isolation and the exposed nature of the solar panels. Power surges, whether caused by lightning or grid fluctuations, can irreparably damage sensitive solar components. Implementing proper lightning and surge protection measures is pivotal to the long-term functionality of your solar setup.

At the crossroads of knowledge and vigilance, comprehending the nature of lightning and power surges is pivotal. Lightning, a captivating yet destructive natural phenomenon, possesses the power to deliver an immense surge of energy within a fraction of a second. Even if a lightning strike doesn't directly hit your cabin, the subsequent electrical currents it induces can travel through conductive paths—such as power lines and cables—threatening irreparable damage to your solar power system. Accompanying this peril is the prospect of power surges, sudden spikes in voltage originating from diverse sources, including utility grid fluctuations and equipment irregularities.

Guardians of Protection: Strategies and Measures

Lightning Rods (Air Terminals)

Erecting lightning rods, also known as air terminals, presents an astute line of defense. Positioned atop or near your cabin, these conductive rods offer an alternative pathway for lightning to travel—safely guiding the energy toward the ground and away from your solar arrays and crucial components.

Surge Protectors (Transient Voltage Suppressors)

Emblematic of vigilance, surge protectors act as sentinels stationed within your system. These guardians identify and intercept voltage spikes and surges, denying them passage to your equipment. Imposing surge protectors at multiple junctures, encompassing the solar panel array, charge controllers, inverters, and battery banks, serves as a comprehensive barrier against the incursion of power anomalies.

Grounding Rituals

A cornerstone of defense, proper grounding acts as a channel for the dissipation of excess energy, guiding it harmlessly into the embrace of the earth. Assemble an intricate network of grounding, encompassing not only your solar panels but also mounting structures and diverse system components. Adherence to electrical code standards is pivotal in the creation of an effective grounding strategy.

Sentinels of Isolation: Transformers

Within the realm of protection, isolation transformers serve as sentinels of isolation. This transformative addition shelters your solar power system from the vagaries of utility grid fluctuations, thus thwarting the propagation of voltage surges through your infrastructure.

The Shield of Quality Equipment

Opting for discerningly chosen, certified equipment offers a layer of natural defense. Many contemporary inverters and charge controllers incorporate innate surge protection features, strengthening the resilience of your system against the tempestuous onslaught of power anomalies.

An Art Form of Installation and Vigilance

The Maestros of Installation

Delving into the realm of lightning protection warrants the counsel of experts. Engaging the expertise of professional electricians or specialists in lightning protection can guide you toward a comprehensive strategy that encapsulates all potential vulnerabilities.

Guardians of Inspection

Pioneering the path of diligence, regular inspections stand as the guardians of your protection measures. Routinely examine the state of your lightning protection system, surge protectors, grounding apparatus, and interconnecting equipment. Vigilance ensures their steadfast functionality.

Voyage of Preemptive Action

For those who dwell in regions prone to lightning storms, a voyage of preemptive action emerges. During severe storms, swiftly disconnecting your solar arrays and suspending your system's operation can prove instrumental in mitigating potential risks.

As custodian of your off-grid solar power system, nurturing a profound understanding of lightning, power surges, and the protective measures enacted is nonpareil. Engage in the quest for enlightenment—comprehend the intricate nuances of lightning protection, surge suppression, and the art of grounding. This voyage equips you with the wisdom to make informed decisions, forging a shield of safety around your cherished investment.

Winterizing Your Off-Grid Cabin and Solar Equipment

As the seasonal tapestry shifts and the world outside your off-grid cabin becomes adorned with frost and snow, the meticulous act of winterizing emerges as a vital chore. Winter, with its formidable challenges and icy embrace, not only tests the comfort of your living space but also poses potential risks to the longevity and functionality of your cherished solar power system.

The Cocoon of Insulation and Sealing

The first brushstroke in the canvas of winterization involves the embrace of insulation. The meticulous sealing of gaps, enveloping walls, roofs, and floors with insulating materials, forms an impervious cocoon that retains the warmth within and repels the cold without. Proper insulation is the cornerstone that stabilizes the internal temperature, reducing the burden on any heating mechanisms you employ.

As winter's chill dances upon the landscape, the choice of a heating system emerges as a pivotal decision. The crackling embrace of a wood-burning stove, the subtle warmth of propane heaters, or the gentle caress of radiant floor heating are all possibilities to be explored. This symphony of warmth, however, requires a conductor—ventilation. Adequate ventilation prevents the stealthy encroachment of moisture and the formation of mold.

Tilt and Snow Removal

The orchestration of solar panels assumes a new melody during winter. Adjusting the tilt of your solar panels can be akin to tuning a musical instrument to produce optimal notes. An inclined position invites the snow to cascade gracefully, ensuring that your panels bask in the spotlight of winter sunlight.

Engaging in the graceful ballet of snow removal becomes a choreography of care. Gently brushing away accumulated snow from the panels, using soft tools to avoid damage, restores their performance to the crescendo of efficiency.

Battery Care and Resilience

The batteries, those repositories of energy, are sensitive to temperature's whims. They react more fervently in colder conditions. Hence, orchestrating a routine of regular battery maintenance becomes the conductor's baton. Monitoring charge states, performing essential upkeep, and even embracing the symphony of a backup generator for extended periods of diminished sunlight orchestrate the harmony of battery resilience.

Antifreeze and Prudent Care

If your cabin boasts the luxury of plumbing, the plumbing elegy resounds with the importance of antifreeze. The delicate choreography of preventing frozen pipes and potential water damage is enacted through draining the plumbing system or infusing it with antifreeze. Toilets, sinks, and drains serenade to the tune of antifreeze, ensuring the rhythm of water's flow remains uninterrupted even as winter's frosty fingers beckon.

In the grand theater of winterization, each element plays a pivotal role—insulation and sealing form the prologue, heating systems compose the melodic interlude, solar panels take center stage with their adjusted tilt and snow-clearing ballet, and the batteries resonate as the underlying harmony. Through it all, prudent plumbing care becomes the poetic epilogue.

BOOK 8

OFF-GRID SOLAR
POWER FOR RVS

CHAPTER 1: INSTALLING SOLAR PANELS ON RVS AND MOTORHOMES

Roof-Mounted vs. Portable Solar Panels: Pros and Cons

In recent years, the surge in environmental awareness and the desire for sustainable living have led to an increasing number of RV and motorhome owners considering solar power as an alternative energy source. This chapter delves into the intricacies of installing solar panels on RVs and motorhomes, exploring the benefits and challenges associated with both roof-mounted and portable solar panels, as well as providing insight into retrofitting these vehicles for solar power and optimizing energy collection during travels.

One of the first considerations for RV and motorhome owners is whether to opt for roof-mounted or portable solar panels. Each option carries its own set of advantages and drawbacks.

Roof-Mounted Solar Panels

Roof-mounted solar panels are a popular choice for those seeking a more permanent and streamlined solution. These panels are affixed directly to the roof of the vehicle, seamlessly integrating with its design. This integration offers aesthetic appeal as the panels become part of the vehicle's exterior. Beyond the visual aspect, roof-mounted panels are also highly efficient since they can be oriented to capture maximum sunlight throughout the day, optimizing energy production.

However, roof-mounted panels do come with certain challenges. The installation process can be more complex and might require professional assistance, adding to the overall cost. Additionally, the fixed orientation of the panels means that they may not be as effective in areas with limited sunlight or during parts of the day when the sun is at an angle. Furthermore, there's a risk of potential damage to the panels due to low-hanging branches or obstructions when driving through certain areas.

Portable Solar Panels

On the other hand, portable solar panels offer a higher degree of flexibility. These panels are not affixed to the vehicle and can be set up at campsites or parking spots to harness solar energy. This versatility allows owners to place the panels in direct sunlight, optimizing energy collection. Portable panels are also relatively easy to install and can be moved around to avoid shading obstacles.

Despite these advantages, portable solar panels have their own set of drawbacks. They need to be set up and taken down each time the vehicle moves, which can be inconvenient. Moreover, their portability makes them susceptible to theft or damage if not properly secured. In terms of aesthetics, they don't offer the seamless integration that roof-mounted panels do.

Retrofitting RVs for Solar Power: Step-by-Step Guide

Retrofitting an RV or motorhome for solar power involves several key steps. This guide aims to provide a comprehensive overview of the process, highlighting important considerations along the way.

Step 1: Energy Audit

Before embarking on the retrofitting journey, it's crucial to conduct an energy audit to determine the energy needs of the vehicle. Consider the appliances and systems that require power, such as lighting, heating, cooling, electronics, and more. This audit will serve as the foundation for calculating the required solar panel capacity.

Step 2: Solar Panel Selection

Choosing the right solar panels depends on factors such as available roof space, energy consumption, and budget. High-efficiency panels might be preferred if space is limited, while budget-conscious owners could opt for more affordable options with slightly lower efficiency. Research and consultation with experts can aid in making an informed decision.

Step 3: Inverter and Battery Bank

An inverter is essential for converting the direct current (DC) produced by solar panels into the alternating current (AC) required by most appliances. A battery bank stores excess energy for use during nighttime or cloudy periods. Selecting an appropriately sized inverter and battery bank is vital to ensure a steady power supply.

Step 4: Mounting and Wiring

For roof-mounted panels, proper mounting is critical to withstand road vibrations and weather conditions. This step might require professional assistance to guarantee a secure and durable installation. Additionally, wiring the panels, inverter, and battery bank together should be executed meticulously to ensure efficient energy transfer.

Step 5: Monitoring System

Integrating a monitoring system allows RV owners to track energy production, consumption, and battery levels. This information empowers them to make adjustments and optimize energy usage.

Step 6: Maintenance and Upkeep

Regular maintenance is key to keeping the solar power system functioning optimally. This involves cleaning the panels, checking for loose connections, and ensuring the battery bank is operating efficiently.

Maximizing Solar Energy Collection While Traveling

The allure of traveling in an RV or motorhome is closely tied to the freedom of the open road. However, this mobility poses challenges when it comes to solar energy collection. To maximize energy generation while on the move, owners can consider the following strategies:

1. Solar Tracking Systems

Solar tracking systems are designed to tilt and rotate solar panels to follow the path of the sun throughout the day. While this technology adds complexity and cost, it significantly boosts energy generation by keeping panels at an optimal angle to the sun.

2. Flexible Solar Panels

Flexible solar panels offer a unique advantage as they can be attached to curved surfaces, potentially expanding the available solar panel area on the vehicle. This can be particularly useful for RVs with unconventional roof designs.

3. Energy-Efficient Appliances

Investing in energy-efficient appliances reduces the overall energy demand of the vehicle. LED lighting, energy-efficient refrigerators, and smart thermostats all contribute to lowering energy consumption.

4. Awnings with Solar Panels

Some RVs come equipped with awnings that incorporate solar panels. These dual-purpose awnings not only provide shade but also generate solar energy. This can be a convenient way to add to the solar capacity without taking up additional roof space.

5. Travel Itinerary Planning

Strategic travel planning can ensure that the RV is parked in optimal positions to receive maximum sunlight during the day. Utilizing apps or tools that provide real-time solar angle information can aid in selecting parking spots.

6. Energy Storage Capacity

Increasing the battery bank's capacity allows for more energy storage, which is especially useful when camping in areas with limited sunlight. This ensures a steady power supply even during cloudy days.

The decision to install solar panels on RVs and motorhomes involves careful consideration of various factors, including panel type, installation method, and energy optimization strategies. Whether choosing roof-mounted or portable panels, the move towards solar power signifies a step towards sustainable and environmentally conscious travel. Retrofitting the vehicle requires meticulous planning and execution, with each step contributing to a robust solar power system. Lastly, maximizing energy collection during travel demands a combination of innovative technologies and mindful energy consumption. With solar power, the open road becomes an even more sustainable and eco-friendly adventure.

CHAPTER 2: BATTERY UPGRADES AND ENERGY STORAGE FOR TRAVELING

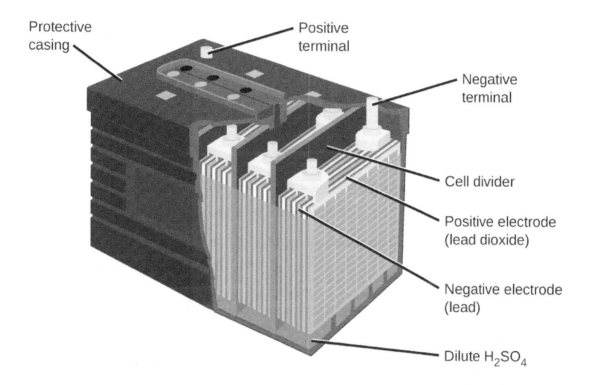

Protective casing

Positive terminal

Negative terminal

Cell divider

Positive electrode (lead dioxide)

Negative electrode (lead)

Dilute H_2SO_4

In the vast realm of recreational vehicles (RVs), where the road becomes home, and adventure intertwines with routine, one aspect stands out as both crucial and ever-evolving: energy storage. As nomadic enthusiasts seek longer journeys and more off-the-grid experiences, the significance of efficient and reliable energy storage solutions has surged to the forefront. This chapter delves into the heart of this matter, exploring the world of battery upgrades for RVs, the potential of solar generators and portable power stations, and indispensable tips for RV battery maintenance and longevity.

Lithium-Ion Battery Upgrades for RVs

The heartbeat of any RV's electrical system lies within its battery bank. Historically, lead-acid batteries dominated this landscape, but in recent times, lithium-ion technology has emerged as a game-changer. Lithium-ion batteries boast a myriad of advantages that resonate profoundly with RV enthusiasts seeking enhanced performance and longevity.

One of the most compelling aspects of lithium-ion batteries is their superior energy density. This property allows them to store more energy in a smaller and lighter package, a characteristic that aligns harmoniously with the spatial constraints of an RV. Moreover, this energy density directly translates into an extended power supply, granting travelers the luxury of prolonged boondocking experiences, where they can detach from traditional campgrounds and embrace the tranquility of nature without worrying about power scarcity.

Additionally, lithium-ion batteries exhibit a notably higher depth of discharge compared to their lead-acid counterparts. This implies that RVers can tap into a larger portion of the battery's capacity without causing damage. Consequently, this translates into a longer-lasting power source and a more reliable energy reservoir for essential appliances such as refrigerators, lighting systems, and communication devices.

The efficiency of lithium-ion batteries also shines through in their charge and discharge rates. Unlike lead-acid batteries, which experience voltage sag and decreased efficiency during rapid discharges, lithium-ion batteries maintain a stable output throughout. This attribute not only accommodates the sudden power demands of appliances like air conditioners but also optimizes energy capture during regenerative processes, such as braking while towing.

Nonetheless, while the advantages of lithium-ion batteries for RVs are undeniable, their adoption presents some considerations. Primarily, cost remains a prominent factor. Lithium-ion batteries tend to have a higher upfront expense compared to traditional lead-acid batteries. However, enthusiasts often view this as a long-term investment due to their extended lifespan and performance benefits.

Moreover, proper battery management systems (BMS) are imperative when integrating lithium-ion batteries. These systems oversee individual cell voltages, temperature, and overall health, ensuring a safe and optimized operation. As lithium-ion technology evolves, BMS units are becoming increasingly sophisticated, offering features such as remote monitoring and automatic cell balancing.

Solar Generators and Portable Power Stations

The marriage of RVs and renewable energy sources has given birth to a dynamic duo: solar generators and portable power stations. In an era of environmental consciousness and amplified wanderlust, these solutions stand as the embodiment of sustainable nomadic living.

Solar generators, equipped with photovoltaic panels, harness the sun's energy and convert it into electricity. Their allure lies in their simplicity and autonomy. By leveraging the power of the sun, RV travelers can recharge their batteries and power appliances without relying on traditional grid connections. This is particularly advantageous for those who venture into remote areas where power outlets are scarce.

The modularity of solar generators also deserves acclaim. RVers can tailor their systems according to their energy needs, expanding the number of panels and storage capacity as required. Additionally, solar generators produce zero emissions and minimal noise, aligning seamlessly with the RV ethos of treading lightly upon the earth.

Portable power stations, on the other hand, encompass a broader spectrum of energy sources. While they can integrate solar panels, they can also be charged through traditional electrical outlets or even the vehicle's alternator. These power stations encapsulate various charging methods, providing a versatile solution for energy storage while on the move.

In recent years, technological strides have significantly enhanced the capabilities of these power stations. Lithium-ion batteries, once again, steal the spotlight here due to their high energy density and quick charge-discharge cycles. This allows travelers to swiftly recharge during short stops and harness power even when sunlight is elusive.

RV Battery Maintenance and Longevity Tips

Irrespective of the chosen energy storage solution, meticulous battery maintenance is pivotal to ensure longevity and optimal performance. RVers can adhere to several cardinal practices to maximize the lifespan of their energy storage systems.

1. **Regular Inspection:** Routinely examine the battery bank for signs of corrosion, leaks, or physical damage. Addressing issues promptly can prevent further deterioration.
2. **Charge Control:** Avoid deep discharges whenever possible, as these can strain the battery and curtail its lifespan. Consider installing a battery monitor to track state-of-charge accurately.
3. **Climate Consideration:** Extreme temperatures can impact battery health. Insulate battery compartments and avoid exposing them to temperature extremes to maintain optimal performance.
4. **Equalization Charges:** If using lead-acid batteries, administer periodic equalization charges to rebalance cell voltages and prevent sulfation, which can hinder capacity.
5. **Ventilation:** Ensure proper ventilation around the battery bank, as off-gassing during charging can lead to the accumulation of explosive hydrogen gas.

6. **Routine Charging:** If storing the RV during off-seasons, implement a routine charging schedule to prevent battery self-discharge and sulfation.

7. **Maintenance of Solar Components:** For solar-powered systems, keep panels clean and free from debris to optimize energy capture efficiency.

8. **BMS Oversight:** For those employing lithium-ion technology, stay vigilant with the battery management system to monitor cell health and prevent overcharging or overheating.

As RV travel evolves beyond the traditional bounds of highways and campgrounds, the energy storage landscape adapts in tandem. Battery upgrades, propelled by the advancements of lithium-ion technology, open doors to extended journeys and enhanced power reliability. Solar generators and portable power stations usher in an era of sustainable exploration, intertwining the thrill of the open road with the responsibility of preserving the planet. With careful battery maintenance practices, RV enthusiasts can unlock the full potential of these systems, embarking on odysseys defined by freedom, resilience, and a profound connection to the ever-changing tapestry of landscapes that stretches before them.

CHAPTER 3: SOLAR-POWERED RV APPLIANCES AND ELECTRONICS

As the world embraces sustainable and eco-friendly alternatives, the realm of recreational vehicles (RVs) has not been left untouched. The allure of hitting the open road, exploring new destinations, and immersing oneself in nature's beauty has led to a surge in RV enthusiasts. However, the traditional mode of RV travel often relies heavily on non-renewable energy sources, which can be both environmentally harmful and economically draining. This is where solar-powered RV appliances and electronics come to the forefront, transforming the way we experience RV living.

Solar-Powered RV Air Conditioners and Fans

One of the primary concerns when embarking on an RV journey, especially in warmer climates, is maintaining a comfortable indoor temperature. Traditional RV air conditioning units are notorious for their high energy consumption, which not only drains the vehicle's fuel or electricity reserves but also contributes to greenhouse gas emissions. This is where solar-powered RV air conditioners enter the scene as a game-changer.

Solar-powered RV air conditioners harness the abundant energy of the sun to cool the interior of the vehicle, minimizing the reliance on traditional power sources. These systems typically consist of photovoltaic (PV) panels mounted on the roof of the RV. These panels capture sunlight and convert it into electricity, which is then used to power the air conditioning unit. This innovative solution not only reduces the carbon footprint associated with RV travel but also provides travelers with a cost-effective way to stay cool during their journeys.

In addition to air conditioning, solar-powered RV fans have also gained popularity. These fans are strategically positioned to ensure proper air circulation within the vehicle. They help maintain a comfortable temperature and reduce the need for continuous air conditioning, further optimizing energy consumption.

Efficient RV Lighting and Entertainment Systems

Creating a cozy and enjoyable ambiance within the RV is an integral part of the travel experience. Traditional incandescent bulbs, though effective, tend to be power-hungry and need frequent replacements. This is where efficient RV lighting solutions come into play, enhancing the interior while conserving energy.

LED (Light Emitting Diode) lighting has revolutionized RV illumination. These lights are designed to consume significantly less energy than conventional bulbs, making them an ideal choice for solar-powered RVs. LED lights are not only energy-efficient but also last much longer, reducing the need for frequent replacements during the journey.

Moreover, solar-powered RVs extend their eco-friendly approach to entertainment systems as well. Entertainment setups, such as flat-screen televisions, audio systems, and even charging stations for electronic devices, can be powered by solar energy. This significantly reduces the need to rely on external power sources or deplete the vehicle's batteries.

Solar-Powered RV Water Heaters and Showers

Access to hot water is a luxury that RV travelers cherish, particularly after a long day of exploration. Traditional RV water heaters often run on propane, which not only adds to the operational costs but also emits greenhouse gases. Solar-powered RV water heaters provide an alternative that aligns with the principles of sustainability.

These water heating systems utilize solar thermal technology. Solar panels, often installed on the roof of the RV, capture sunlight and transfer its heat to a fluid that circulates within the system. This heated fluid, in turn, warms up the water in the storage tank. This ingenious process ensures a consistent supply of hot water while significantly reducing the reliance on traditional energy sources.

Solar-powered RV showers complement the water heating systems by providing a comfortable and eco-friendly bathing experience. These showers are designed with water-saving features and can be equipped with temperature controls to ensure a relaxing experience without wastage.

The evolution of solar-powered RV appliances and electronics marks a pivotal moment in the world of recreational travel. By harnessing the power of the sun, RV enthusiasts can now embark on journeys that are not only thrilling and adventurous but also aligned with sustainable living practices. Solar-powered air conditioners keep the interior cool without compromising the environment, efficient lighting and entertainment systems enhance the on-road experience, and solar-powered water heaters and showers provide the comfort of home while minimizing the carbon footprint.

CHAPTER 4: SOLAR RV MAINTENANCE AND TIPS FOR LIFE ON THE ROAD

The allure of the open road, the freedom to explore new horizons, and the ability to take your home wherever you go - are the driving forces behind the growing trend of living life on the road in a recreational vehicle (RV). As more individuals and families embrace this nomadic lifestyle, harnessing the power of solar energy has emerged as a sustainable and practical solution to meet energy needs while on the go. In this chapter, we delve into the crucial aspects of solar RV maintenance and provide invaluable tips for maximizing your solar-powered journey.

Cleaning and Maintaining Solar Panels on the Go

Imagine waking up to a breathtaking sunrise in a remote camping spot, surrounded by nature's beauty, and knowing that your solar panels have been efficiently gathering energy from the sun while you slept. To ensure this scenario becomes a reality, understanding how to properly clean and maintain your RV's solar panels is paramount.

Solar panels are the heart of your RV's energy system. They convert sunlight into electricity, which powers everything from your lights to your refrigerator. Keeping them clean and functioning optimally is essential for a smooth and energy-efficient journey.

1. **Regular Cleaning**: Just like any other surface, solar panels accumulate dirt, dust, and debris over time. Regular cleaning, ideally once every few weeks or as needed, will ensure that the panels receive maximum sunlight exposure. Use a soft brush or cloth to gently remove dirt and grime. Avoid using abrasive materials that could scratch the surface.

2. **Water Rinse**: Before cleaning the panels, give them a gentle water rinse to loosen dirt and particles. This step prevents unnecessary friction during the cleaning process.

3. **Mild Detergent**: For tougher stains, a mixture of mild detergent and water can be used. However, make sure to rinse thoroughly to prevent any residue that might hinder the panel's efficiency.

4. **Avoid Abrasives**: Abrasive cleaners, harsh chemicals, and rough materials should be avoided at all costs. They can damage the panel's protective coating and reduce its effectiveness.

5. **Inspect for Damage**: While cleaning, take the opportunity to inspect the panels for any signs of damage, such as cracks or hot spots. Addressing such issues promptly can prevent further complications down the road.

Planning Your Routes Based on Solar Energy Availability

One of the most liberating aspects of RV travel is the ability to choose your own path. However, to make the most of your solar energy system, it's wise to plan your routes with solar energy availability in mind. Here's how:

1. **Sun Angle and Direction**: Solar panels are most efficient when they are angled towards the sun. This means that if you're traveling north or south, you'll want to park your RV with the panels facing east in the morning and west in the afternoon. If you're traveling east or west, the opposite applies.

2. **Research Campsites**: Thanks to technology, you can now research campsites online and even find reviews from other RVers. Look for sites with open, unobstructed views of the sky to ensure maximum sunlight exposure.

3. **Weather Considerations**: While sunny days are ideal, it's important to consider cloudy days as well. Investing in a solar charge controller with MPPT (Maximum Power Point Tracking) technology can optimize energy production even on cloudy days.

4. **Battery Storage**: Another key factor is your RV's battery storage capacity. If you plan to spend a significant amount of time in one location, ensure your battery bank is large enough to store excess energy generated during sunny days for use during nights or cloudy periods.

Staying Connected: Finding RV-Friendly Solar Campsites

In the realm of RV travel, not all campsites are created equal when it comes to accommodating solar-powered vehicles. However, an increasing number of RV-friendly campsites are recognizing the appeal of solar-powered guests. Here's how to find them:

1. **Online Directories**: Numerous online directories and platforms cater to the RV community. These platforms allow you to filter campsites based on their amenities, including solar hookups. Websites like RV Trip Wizard, Campendium, and AllStays provide valuable information.

2. **Community Forums**: Engaging with fellow RVers through online forums and social media groups can yield firsthand recommendations for solar-friendly campsites. These communities often share hidden gems that might not be listed in conventional directories.

3. **Contact Campsite Management**: If you find a campsite you're interested in, don't hesitate to reach out to the management. Inquiring about their electrical amenities and mentioning your reliance on solar power can help you gauge whether they can accommodate your needs.

4. **National and State Parks**: Many national and state parks are embracing solar technology. While not all may have dedicated solar hookups, some offer primitive camping options that allow you to harness solar power without the need for traditional electrical hookups.

The marriage of RV travel and solar energy opens up a world of possibilities for adventurers seeking sustainable and independent living. By understanding the nuances of solar panel maintenance, strategically planning your routes, and seeking out solar-friendly campsites, you can embark on a solar-powered journey that combines the best of nature and modern technology. As you navigate the open road, remember that while the destination is important, it's the journey - powered by the sun - that truly makes the experience unforgettable.

BOOK 9

OFF-GRID SOLAR POWER FOR BOATS

CHAPTER 1: MARINE SOLAR PANEL INSTALLATION AND SETUP

The serene beauty of the open sea has always captivated humanity, and for many, the allure of sailing represents the ultimate freedom. However, modern sailing isn't just about harnessing the wind; it's also about integrating cutting-edge technology to make the experience more comfortable, sustainable, and efficient. Among these technologies, marine solar panels have emerged as a game-changer, providing boats with a renewable source of energy to power various onboard systems. In this chapter, we will delve into the intricacies of marine solar panel installation and setup, exploring boat-specific mounting solutions, the utility of flexible and lightweight panels, and optimal solar panel placement on sailboats and motor yachts.

Boat-Specific Solar Panel Mounting Solutions

When the voyage of renewable energy meets the vast expanse of the open sea, a unique chapter in the world of solar power unfolds—boat-specific solar panel mounting solutions. Unlike their land-bound counterparts, boats face a dynamic and ever-changing environment, demanding innovative approaches to solar panel installation that navigate challenges such as saltwater exposure, limited space, and constant movement.

Fixed-Mount Systems

Anchored firmly to the boat's deck or roof, fixed-mount solar panels provide a stable platform for harnessing the sun's energy. This approach is well-suited for vessels with ample roof space and minimal shading concerns. However, the decision to opt for a fixed-mount system should be made with consideration for the boat's structure, as irregular deck shapes or a desire to preserve the boat's original appearance might present challenges.

Adjustable Tilt Systems

The sun, a celestial compass, moves across the sky with the passage of time. In response, adjustable tilt systems embrace adaptability. By enabling solar panels to be tilted toward the sun's angle, these mounts optimize energy capture, especially during periods of low-light conditions. Boats that chart diverse courses or traverse latitudes with varying sun angles can benefit from the flexibility of adjustable tilt systems, ensuring a consistent flow of energy throughout the journey.

Pole Mounts

As a ship's mast reaches towards the heavens, pole mounts extend solar panels above the deck's expanse. These elevated panels are perched upon sturdy poles, either positioned directly on the deck or attached to the boat's railings. For vessels grappling with limited deck space, pole mounts provide a pragmatic solution. By lifting panels off the deck, more room is liberated, reducing shading issues that could impede energy production. Yet, the embrace of pole mounts is a dance of balance, as the elevated structure introduces wind resistance and can influence the boat's aesthetic composition.

Each mounting solution paints a unique canvas upon the boat's surface, shaping its interaction with the sun's radiant embrace. Fixed-mount systems ground solar panels with unwavering determination, while adjustable tilt systems pivot in sync with the sun's arc. In contrast, pole mounts elevate panels like banners of energy defiance against the backdrop of the open sky.

Flexible and Lightweight Solar Panels for Boats

In the maritime realm, where space is a premium and the demand for efficient energy solutions is paramount, the emergence of flexible and lightweight solar panels has revolutionized the way boats harness the power of the sun. These innovative solar panels, designed specifically for nautical environments, introduce a new chapter in maritime sustainability by addressing the unique challenges and opportunities presented by life on the water.

A Seafaring Evolution

Gone are the days of rigid, bulky solar panels that required extensive installation efforts and limited placement options. The advent of flexible and lightweight solar panels has ushered in a new era of adaptability and convenience for boat owners seeking to integrate solar energy into their seafaring lifestyle. These panels are engineered with materials that allow them to curve and flex, conforming to the contours of boat surfaces with a grace that was previously unattainable.

Conquering Space Constraints

One of the most pressing challenges in marine solar installation is the scarcity of available space. Boats, often designed with streamlined dimensions to navigate waterways, lack the expansive rooftops of land-based structures. Flexible and lightweight solar panels rise to this challenge by offering versatile placement options. These panels can be affixed to curved surfaces, like the arches of a sailboat or the contours of a cabin roof, maximizing the use of available space and transforming previously untapped areas into efficient energy-generating zones.

Navigating the Waters of Efficiency

Efficiency is a guiding star in the maritime realm. Flexible and lightweight solar panels are equipped with cutting-edge photovoltaic technology that converts sunlight into electricity with impressive efficiency. This efficiency is not only a boon for maintaining power-hungry onboard systems but also a testament to the panels' adaptability. Even when partially shaded or exposed to indirect sunlight, these panels continue to contribute to a boat's energy needs, ensuring a consistent power supply throughout a voyage.

Durability in the Face of Elements

Boats navigate through an environment that demands durability and resilience. Flexible and lightweight solar panels rise to this challenge by incorporating rugged materials that withstand the rigors of marine life. They are designed to resist corrosion from saltwater exposure, endure harsh weather conditions, and even withstand the impact of footsteps, ensuring that they remain operational and reliable even in the most demanding situations.

The Aesthetic Symphony

Beyond their functional prowess, flexible and lightweight solar panels also compose an aesthetic symphony. Their ability to seamlessly integrate with a boat's design enhances the vessel's visual appeal, without disrupting its lines or profile. These panels become a harmonious addition, accentuating the boat's elegance while contributing to its self-sufficiency.

Charting a Sustainable Course

In the grand narrative of maritime sustainability, flexible and lightweight solar panels chart a transformative course. They empower boat owners to embrace clean energy solutions that not only reduce reliance on traditional fuel sources but also minimize the ecological footprint of seafaring adventures. As boats glide across the water, the panels silently and efficiently convert sunlight into power, propelling vessels toward a more sustainable and environmentally conscious future.

Solar Panel Placement on Sailboats and Motor Yachts

The strategic placement of solar panels is essential to harnessing the maximum amount of sunlight and reaping the full benefits of solar energy. The layout and orientation of panels can vary between sailboats and motor yachts due to differences in deck space and energy requirements.

Sailboats:

Sailboats often have limited deck space, which necessitates thoughtful planning for solar panel placement:

- **Cabin Top:** Many sailboats opt to install solar panels on the cabin top, as this area is generally flat and unobstructed. Panels can be mounted using adjustable tilt systems to capture sunlight from various angles.
- **Bimini or Dodger:** Flexible solar panels can be integrated into the bimini or dodger, providing shade while generating energy. This dual-purpose setup is particularly advantageous for boats navigating sunny climates.

- **Mast or Boom:** Some innovative designs involve mounting panels on the mast or boom. However, this approach requires careful engineering to ensure panels do not interfere with the boat's rigging or obstruct sail handling.

Motor Yachts:

Motor yachts typically have more deck space, offering greater flexibility in panel placement:

- **Flybridge:** The expansive flybridge area on motor yachts provides an excellent location for solar panels. Panels can be fixed to the hardtop or integrated into sun loungers, offering dual functionality.
- **Aft Deck:** Solar panels can be seamlessly integrated into the design of the aft deck. This area is easily accessible and can accommodate both rigid and flexible panels.
- **Rails and Guardrails:** Flexible panels can be attached to the boat's railing or guardrails, effectively utilizing vertical space. This is a practical solution for motor yachts with limited horizontal deck space.

As the maritime industry embraces sustainable practices, marine solar panels have emerged as a pivotal technology for modern boats. With boat-specific mounting solutions catering to diverse needs, the advent of flexible and lightweight panels, and a deep understanding of optimal panel placement, boat owners now have the tools to seamlessly integrate solar energy into their sailing experience. The synergy of innovation and tradition is steering the course toward a greener, more efficient future for the world of boating.

CHAPTER 2: BATTERY BANKS AND CHARGING FOR BOATS

In the realm of boating, where freedom meets the open water, reliable power sources are imperative to ensure a safe and enjoyable journey. Battery banks, as the lifeblood of a boat's electrical systems, hold a pivotal role in this regard. Moreover, with the increasing concern for environmental sustainability, integrating renewable energy sources such as solar, wind, and hydro turbines has become an enticing option. In this chapter, we delve into the intricate world of battery bank configuration and placement, the significance of solar charge controllers in marine environments, and the potential of wind and hydro turbines as supplementary charging sources for boats.

Marine Battery Bank Configuration and Placement

At the core of any marine vessel's electrical system lies the battery bank - an assortment of batteries wired together to amass and store electrical energy. The selection of the battery bank's configuration and its optimal placement is a decision that significantly influences a boat's performance, longevity, and overall reliability.

Battery Configuration

The configuration of a battery bank plays a substantial role in determining the voltage and capacity of the system. Boats often employ two main types of batteries: deep-cycle and starting batteries. Deep-cycle batteries are designed to provide a steady amount of power over a prolonged period, making them ideal for auxiliary systems such as lighting, refrigeration, and onboard electronics. Starting batteries, on the other hand, are designed to provide short bursts of high current to start the engine.

Several configurations are commonly employed, such as series, parallel, and series-parallel. Series connections increase the voltage while maintaining the battery capacity. Parallel connections boost the capacity while maintaining the voltage. Series-parallel configurations offer a balance between increased voltage and capacity, making them well-suited for boats requiring both steady auxiliary power and reliable engine starting capability.

Placement Considerations

The physical placement of the battery bank within the boat is crucial for maintaining balance, stability, and safety. Batteries are weighty components, and improper placement can affect a boat's stability and maneuverability. Generally, the battery bank should be situated near the boat's center of gravity, usually below the waterline, to ensure optimal weight distribution.

Additionally, adequate ventilation is paramount to prevent overheating and the buildup of potentially hazardous gases emitted during charging. Battery boxes or compartments with proper ventilation systems are essential to avoid safety hazards and to prolong the batteries' lifespan.

Solar Charge Controllers for Marine Environments

In the quest for sustainable and efficient energy sources, solar power has emerged as a promising option for boaters. Solar panels, equipped with photovoltaic cells, harness sunlight and convert it into electrical energy. However, the fluctuating nature of solar energy production necessitates the use of solar charge controllers, especially in marine environments where conditions can be harsh and unpredictable.

Role of Solar Charge Controllers

Solar charge controllers, also known as solar regulators, act as intermediaries between solar panels and battery banks. Their primary function is to regulate the voltage and current reaching the batteries, preventing overcharging and deep discharge, which can significantly reduce battery life.

In marine environments, where exposure to moisture, salt, and temperature variations is commonplace, selecting a charge controller with a suitable ingress protection (IP) rating is essential. An IP rating indicates the controller's resistance to dust and water infiltration, ensuring its longevity and reliability under maritime conditions.

Types of Solar Charge Controllers

Two main types of solar charge controllers are PWM (Pulse-Width Modulation) and MPPT (Maximum Power Point Tracking). PWM controllers are cost-effective and regulate the charging process by intermittently interrupting the flow of energy to the batteries. MPPT controllers, however, are more sophisticated and efficient. They track the maximum power point of the solar panels, allowing them to deliver the optimal amount of energy to the batteries even during varying weather conditions. While PWM controllers suffice for basic setups, MPPT controllers are more suitable for boats with larger solar arrays and higher energy demands.

Integration Challenges

Integrating solar power into marine environments presents its set of challenges. Ensuring the proper angle and orientation of solar panels to maximize sunlight exposure is essential. Moreover, saltwater corrosion and the constant motion of the boat necessitate durable mounting solutions and careful wiring to mitigate wear and tear.

Wind and Hydro Turbines as Supplementary Charging Sources

While solar power is a popular choice, boaters are increasingly exploring supplementary charging sources like wind and hydro turbines to diversify their energy generation and ensure a consistent power supply even when sunlight is limited.

Wind Turbines

Wind turbines harness the kinetic energy of the wind and convert it into rotational energy, which is then transformed into electricity. In marine environments, wind turbines can be mounted on masts or other elevated structures to capture the prevailing winds effectively. However, wind power comes with challenges such as noise generation, potential interference with navigation equipment, and the need for proper balancing to avoid undue stress on the boat's structure.

Hydro Turbines

Hydro turbines, or water turbines, leverage the flow of water to generate mechanical energy, which is then converted into electricity. In the context of boating, hydro turbines can be installed in areas with strong currents, such as tidal zones or rivers. They offer a consistent power source, but installation requires careful consideration of the turbine's size and design to ensure it withstands underwater conditions and remains unobtrusive to marine life.

Integrating Turbines

Incorporating wind and hydro turbines into a boat's energy ecosystem requires thorough planning and adaptation. Turbines need to be properly secured to withstand the harsh marine environment, and their electrical systems must be integrated with the battery bank and other charging sources. Monitoring systems that provide real-time information about turbine performance and power generation are crucial to optimizing their efficiency.

Battery banks and charging systems are the lifelines of modern boats, powering everything from navigation equipment to creature comforts. Careful consideration of battery configuration and placement can ensure optimal performance and safety. Solar charge controllers play a critical role in regulating solar energy, especially in challenging marine environments. Exploring wind and hydro turbines as supplementary charging sources adds a layer of reliability and sustainability, albeit with unique challenges. As boating technology continues to evolve, the integration of these systems becomes not just a matter of convenience but a crucial step toward a more efficient, eco-friendly, and enjoyable boating experience.

CHAPTER 3: SOLAR-POWERED BOAT A PPLIANCES AND NAVIGATION SYSTEMS

In the realm of maritime innovation, where concerns about sustainability and environmental impact loom large, the integration of solar power has emerged as a promising avenue for transforming the way boats function. The allure of harnessing the sun's energy to power boat appliances and navigation systems not only holds the potential to reduce carbon emissions but also enhances the autonomy and efficiency of waterborne vessels. This chapter delves into the intriguing domain of solar-powered boat appliances and navigation systems, exploring their implications for sustainable boating practices and the seamless navigation of waterways.

Solar-Powered Boat Refrigeration and Cooking

Imagine a scenario where a boat's refrigeration and cooking systems can function without drawing power from traditional sources like generators or batteries. This is not just a futuristic fantasy but a practical reality being achieved through the integration of solar power. The innovation lies in the installation of solar panels on the boat's surface, which capture sunlight and convert it into electricity, powering the various appliances onboard.

Solar-Powered Refrigeration

Traditional refrigeration systems on boats have historically depended on non-renewable sources of energy, often leading to fuel consumption and emissions. The introduction of solar-powered refrigeration marks a significant departure from this norm. These systems utilize highly efficient solar panels to generate electricity, which, in turn, powers the refrigeration unit. The solar panels can be integrated into the boat's structure or mounted on its roof, ensuring maximum exposure to sunlight.

The benefits of solar-powered refrigeration are manifold. It not only reduces the ecological footprint of the boat but also offers a consistent and reliable cooling solution, even in remote locations. This becomes particularly important for extended journeys where provisioning and food storage are critical. Moreover, solar-powered refrigeration systems often come with intelligent energy management features, optimizing power consumption and storage, thereby enhancing the overall efficiency of the boat's energy usage.

Solar-Powered Cooking

Just as refrigeration systems have transitioned to solar power, cooking appliances onboard boats are also undergoing a similar transformation. Solar-powered cooking solutions encompass a range of devices such as stoves, ovens, and grills that operate using electricity generated from solar panels. These panels can either directly power electric cooking devices or charge batteries that store energy for later use.

The advantages of solar-powered cooking are multifold. They eliminate the need for traditional fuel sources like propane or charcoal, which are not only resource-intensive but can also pose safety risks on boats. Solar-powered cooking is cleaner, quieter, and eliminates emissions. Additionally, these systems are highly adaptable and can be customized to suit the specific needs of different types of boats, whether they are leisure yachts or commercial vessels.

However, challenges do exist. The efficacy of solar-powered cooking depends on factors like the amount of sunlight available, the energy efficiency of the cooking appliances, and the capacity of the energy storage system. While advancements in solar panel technology and energy storage solutions are mitigating these challenges, it's essential to strike a balance between energy consumption and supply for seamless cooking operations.

Solar-Powered GPS and Communication Devices

The reliance on accurate navigation and efficient communication is paramount in the maritime domain. Traditionally, GPS and communication devices on boats have relied on power sources such as batteries, which, if depleted, could compromise the safety and functionality of the vessel. The integration of solar power into these systems has revolutionized the way boats navigate and communicate, offering enhanced reliability and sustainability.

Solar-Powered GPS

Global Positioning System (GPS) technology is the backbone of modern maritime navigation. The integration of solar power into GPS devices ensures a continuous and reliable power supply, thereby eliminating concerns about battery depletion during extended journeys. Solar panels can be strategically positioned on the boat's deck or superstructure to capture sunlight optimally.

One of the most significant advantages of solar-powered GPS is the reduction of downtime caused by battery changes or recharging. This is especially crucial in critical situations where accurate navigation can be a matter of life and death. Moreover, solar-powered GPS contributes to sustainability by reducing the reliance on disposable batteries or fuel-powered generators, thereby curbing emissions and minimizing the ecological impact.

Solar-Powered Communication Devices

Effective communication is essential for the safety and coordination of activities on boats. Communication devices like radios, satellite phones, and emergency beacons play a pivotal role in ensuring seamless communication with other vessels, shore stations, or emergency services. Integrating solar power into these devices ensures their operability throughout the journey.

Solar-powered communication devices provide a reliable and sustainable solution, reducing the risk of communication breakdown due to power shortages. In remote or off-grid areas where traditional power sources are scarce, solar-powered communication devices offer a lifeline, enabling boats to stay connected even in challenging conditions. Additionally, the reduced reliance on non-renewable power sources contributes to the overall eco-friendliness of boating practices.

LED Navigation Lights and Solar Dock Lighting

Safety is paramount in maritime travel, and navigation lights play a crucial role in preventing collisions and ensuring safe passage. Traditionally, these lights have drawn power from the boat's electrical system, leading to energy consumption. Solar-powered LED navigation lights have emerged as an energy-efficient and sustainable alternative.

Solar-Powered LED Navigation Lights

LED (Light Emitting Diode) technology has transformed lighting solutions across various industries due to its energy efficiency and longevity. When integrated into navigation lights, LEDs reduce power consumption while providing bright and highly visible signals to other vessels. The incorporation of solar panels to power these LEDs further enhances their efficiency.

Solar-powered LED navigation lights offer several benefits. Firstly, they significantly extend the lifespan of onboard batteries by minimizing their use for lighting purposes. This, in turn, reduces the frequency of battery replacements and disposal, contributing to a greener maritime ecosystem. Secondly, solar-powered LED lights are especially valuable in situations where boats are anchored or moored for extended periods, as they can operate autonomously without depleting the boat's energy reserves.

Solar Dock Lighting

Dock lighting is essential for safe and secure docking procedures, particularly during nighttime or low-visibility conditions. Solar-powered dock lighting systems utilize solar panels to capture sunlight during the day and store energy for illuminating the dock area at night. These systems often consist of embedded LEDs along the dock edges or solar-powered bollard lights.

The advantage of solar dock lighting is twofold. It enhances the safety of docking maneuvers by providing clear visibility, and it does so without adding to the boat's power consumption. This is especially useful in marinas or docking facilities where connecting to shore power might be inconvenient or unavailable. Furthermore, solar dock lighting systems contribute to the aesthetics of the marina environment while promoting sustainable practices.

The incorporation of solar power into boat appliances and navigation systems represents a paradigm shift in maritime technology and practices. From refrigeration and cooking to GPS devices and communication equipment, as well as LED navigation lights and dock lighting, solar power has the potential to revolutionize how boats operate. By reducing reliance on non-renewable energy sources, these innovations contribute to environmental preservation while enhancing safety, efficiency, and autonomy in waterborne journeys. As solar panel technology continues to advance, the integration of solar power into boats is poised to become even more seamless and widespread, propelling the maritime industry toward a greener and more sustainable future.

CHAPTER 4: OFF-GRID LIVING AFLOAT: SAFETY AND MAINTENANCE

In the vast tapestry of human existence, the pursuit of a simpler, self-sustaining lifestyle has rekindled interest in off-grid living. While such a lifestyle can take many forms, one particularly captivating variation is the idea of living afloat on watercraft. Picture this: a life unshackled from the confines of land, where the soothing rhythm of water replaces the cacophony of urban life. Yet, beneath the allure of this romantic notion lies a pragmatic concern—safety and maintenance.

Marine Safety Equipment and Emergency Preparedness

The adventure of off-grid living on water holds immense allure, but it is not without its fair share of challenges, and safety is paramount. In this watery world, where the horizon is not bound by walls, the necessity of marine safety equipment and emergency preparedness cannot be overstated.

The heartbeat of marine safety lies in navigational equipment. GPS systems, radar, depth sounders, and electronic charts become the eyes that pierce through mist and darkness. They are the digital companions that aid in steering clear of treacherous waters and ensuring a safe passage. A GPS (Global Positioning System) is no less than a modern-day sextant, allowing mariners to pinpoint their location with extraordinary accuracy.

Just as a painter has a palette of colors, an off-grid mariner has an assortment of personal flotation devices (PFDs). These unassuming life jackets come in various designs tailored for specific purposes. From simple PFDs to intricate inflatable ones, they ensure that, even in the face of adversity, the water's embrace does not turn fatal.

In a world where isolation might be sought but not necessarily welcomed, emergency signaling devices bridge the gap between isolation and assistance. Flares, distress signals, and VHF radios become modern-day smoke signals, casting a plea for help across the waves.

Even on water, the danger of fire is omnipresent. Adequate fire safety measures become the metaphorical fire extinguisher for the heart and hearth of a waterborne dwelling. Fire extinguishers, flame-resistant materials, and a keen understanding of fire prevention techniques form the first line of defense against this fiery adversary.

The stark reality of living afloat is that sometimes, despite all precautions, one might need to abandon ship. Having clear, well-practiced abandon ship protocols can make the difference between life and death. This includes having grab bags—prepared kits containing essentials like water, food, and communication devices—ready to go at a moment's notice.

A cornerstone of maritime safety is staying informed about the weather. Technology grants mariners the ability to monitor changing conditions and predict impending storms. Weather apps, satellite communication systems, and weather stations are the modern-day oracles that guide sailors away from the wrath of nature's tempests.

Saltwater Corrosion Prevention and Rust Control

While the lure of afloat living is rooted in the sense of freedom it offers, there is an ever-present adversary lurking beneath the waves—corrosion. Saltwater, with its corrosive embrace, poses a formidable threat to the structural integrity of any watercraft. Hence, meticulous saltwater corrosion prevention and rust control have become the unsung heroes of this aquatic lifestyle.

Preventing corrosion is akin to a game of strategic sacrifice. Sacrificial anodes, often made of zinc or aluminum, willingly corrode to protect the more valuable metal components of a vessel. These sacrificial lambs are strategically placed around the watercraft, ensuring that the sacrificial process occurs where it causes the least harm.

In the realm of waterborne living, the hull of a vessel acts as both armor and gateway. Applying protective coatings, such as anti-fouling paints, prevents the accumulation of barnacles, algae, and other marine organisms that can accelerate corrosion. These coatings are the invisible shield that stands guard against the relentless assault of saltwater.

For those seeking a more high-tech solution to corrosion, cathodic protection systems are a marvel. These systems use a controlled flow of electrical current to counteract the corrosive effects of saltwater. By conducting a symphony of electrons, these systems ensure that rust remains a distant threat.

In the ever-evolving battle against corrosion, routine inspection and maintenance are the watchmen at the gate. Regularly assessing the condition of the hull, checking sacrificial anodes, and attending to any signs of rust are the rituals that ensure a watercraft's longevity. This vigilance, though demanding, becomes the price paid for a life uninterrupted by the ravages of corrosion.

Maintaining Solar Panels and Equipment on Watercraft

Off-grid living, whether on land or water, shares a common thread—the reliance on renewable energy. Solar panels, the jewel in the crown of sustainable living, adorn many off-grid watercraft. But, like any treasure, they require careful maintenance to continue shining brightly.

The ocean's spray, while refreshing, leaves behind a salty residue that can diminish the efficiency of solar panels. Regular cleaning becomes a ritual of devotion to these energy-providing gods. A gentle touch, using mild detergents and soft brushes, clears away the salt's filmy veil, allowing the panels to drink in the sun's rays more effectively.

Beneath the shimmering surface, the heart of solar energy lies in wiring and connections. Corrosion, the ubiquitous adversary, lurks here as well. Regularly inspecting and cleaning connectors, ensuring a snug fit, and replacing damaged wiring can be the difference between a life bathed in light and one cast into darkness.

In the aquatic world, where space is often at a premium, shade can become a stealthy thief, robbing solar panels of their vitality. Prudent positioning of panels, perhaps with tilting or adjustable mounts, ensures that they bask in the sun's embrace without interruption. A little ingenuity can go a long way in harnessing nature's bounty.

Solar panels might be the sun's gift, but batteries are their treasured keepsakes. Maintaining these energy storehouses is paramount. Regular checks on battery health, ensuring optimal charge levels, and protecting them from extreme temperatures are the pillars upon which an off-grid mariner's power reliance stands.

Just as humans visit doctors for periodic check-ups, solar systems benefit from professional inspections. Calling upon the expertise of solar technicians can unveil hidden issues and ensure that these energy-harvesting mechanisms are operating at their peak efficiency.

Off-grid living afloat is a harmonious marriage of self-sufficiency and aquatic serenity. The allure of gazing at starlit skies from the deck, being rocked to sleep by the lullaby of waves, and witnessing the ever-changing tapestry of ocean life is undeniable. However, beneath this romantic facade lies the reality that safety and maintenance are the guardians of this dream.

Navigational aids, personal flotation devices, and emergency signaling devices ensure survival in times of peril. The relentless battle against saltwater corrosion is waged through strategic anodes, protective coatings, and cathodic protection systems. The luminance of solar panels is maintained through diligent cleaning, meticulous wiring upkeep, and battery care. Embracing off-grid living afloat demands a symbiotic relationship with the elements—a dance where human ingenuity and nature's majesty intertwine.

In this aqueous realm, where horizons stretch beyond sight, and possibilities mirror the endless expanse, the synergy of preparedness and preservation becomes the compass guiding the voyage. As we cast off the moorings of conventional living, we embark on a journey where safety and maintenance carve the path to an enchanting existence on the water's surface.

BOOK 10

ADVANCED OFF-GRID
SOLAR POWER CONCEPTS

CHAPTER 1: HYBRID SOLAR SYSTEMS: INCORPORATING OTHER ENERGY SOURCES

I n the ever-evolving landscape of renewable energy, the quest for efficient and sustainable power generation methods has led to the development of hybrid solar systems. These systems ingeniously combine solar energy with other renewable sources, not only enhancing the reliability of power supply but also contributing significantly to the global transition towards clean energy. This chapter delves into the fascinating realm of hybrid solar systems, exploring their various facets and the benefits they offer in the pursuit of a greener future.

Integrating Wind Power with Solar for Increased Reliability

As the sun bathes the Earth in its abundant energy, and the wind sweeps across landscapes with its kinetic force, combining these two formidable sources of power seems only logical. Hybrid solar wind systems are designed to harness the complementary nature of solar and wind energy. While solar panels thrive under clear skies, wind turbines tend to perform better during cloudy periods or at night when solar output is reduced. By merging these two sources, the system's power generation becomes more consistent and reliable.

Integrating solar and wind components within a single system requires careful consideration of technical challenges. One such challenge is the fluctuating nature of both solar irradiance and wind speed. Engineers and researchers have tackled this issue by implementing advanced control systems that manage the distribution of power based on real-time conditions. Additionally, hybrid systems often employ energy storage solutions, such as batteries, to store excess energy generated during optimal conditions for use during periods of lower energy generation.

The advantages of integrating solar and wind power are manifold. Firstly, this hybrid approach increases the overall capacity factor of the system, leading to a more consistent and stable power supply. Secondly, it allows for better utilization of resources, as the system can generate power from both sources simultaneously or individually, depending on the prevailing weather conditions. Finally, hybrid systems demonstrate an increased potential to deliver power when it's needed the most, which is a critical factor in ensuring a reliable energy supply.

Hybrid Solar Generators: Biomass and Micro-Hydro

Biomass energy, derived from organic materials, has long been a reliable source of power in various forms. When combined with solar energy, the result is a potent hybrid system that offers benefits beyond what each source can provide individually. Biomass can be burned to generate heat, which in turn can be converted into electricity. By coupling this process with solar power, the overall efficiency of the system increases significantly.

Micro-hydroelectric systems, which harness the energy from flowing water, have proven to be an effective means of generating electricity, especially in areas with access to streams or rivers. When integrated into hybrid solar setups, micro-hydro systems can function as a supplementary power source during times when solar energy generation might be limited. This combination ensures a more continuous energy supply, reducing dependency on a single source.

While the hybridization of solar energy with biomass and micro-hydro systems offers promising advantages, it's crucial to address potential environmental and resource-related concerns. The sustainability of biomass sources, for instance, requires diligent management to prevent deforestation or depletion of natural resources. Similarly, the ecological impact of micro-hydro installations on aquatic ecosystems necessitates careful planning and adherence to environmental regulations.

Grid-Interactive Systems and Net Metering

In an era where consumers are becoming increasingly conscious of their energy consumption patterns, grid-interactive hybrid solar systems have emerged as a means of active participation. These systems enable users to not only generate their own power but also interact with the grid. Excess energy generated by the system can be fed back into the grid, earning consumers credits or payments, depending on the net metering policies in place.

Net metering, a cornerstone of grid-interactive systems, allows consumers to strike a balance between their energy generation and consumption. When the hybrid solar system generates surplus energy, it is fed into the grid, effectively "spinning the meter" backward. During periods when the system's energy generation is insufficient, such as at night, consumers draw energy from the grid. The net energy consumed or generated determines the final billing or credit received.

While the concept of net metering holds significant promise, its implementation faces certain challenges. Regulatory frameworks and policies vary from region to region, impacting the feasibility and benefits of net metering. Moreover, technical aspects such as metering accuracy and grid compatibility need to be meticulously addressed to ensure smooth integration and fair compensation.

Hybrid solar systems that incorporate other energy sources represent a remarkable stride in the pursuit of a sustainable energy future. By synergizing the strengths of solar power with wind, biomass, and micro-hydro systems, these hybrid configurations offer enhanced reliability, stability, and environmental benefits. The integration of grid-interactive systems and net metering further empowers consumers to actively engage in the energy ecosystem while contributing to the reduction of carbon footprints on a global scale. As technology continues to advance and renewable energy solutions evolve, hybrid solar systems stand as a testament to human ingenuity and our commitment to forging a cleaner and greener world.

CHAPTER 2: OFF-GRID SOLAR FOR LARGER APPLICATIONS AND HOMESTEADS

In an era characterized by growing environmental concerns and a shift towards sustainable energy sources, off-grid solar systems have emerged as a practical and environmentally friendly solution. They offer an alternative to conventional power sources, particularly in remote areas or homesteads where grid connectivity is either challenging or non-existent. This chapter delves into the intricate details of harnessing solar energy for larger applications, such as farms and homesteads, where energy demands are higher and more diversified.

Designing Off-Grid Solar for Farms and Homesteads

The design of off-grid solar systems for farms and homesteads requires a comprehensive understanding of energy needs, solar potential, and efficient system configuration. Unlike residential setups, these applications demand a more intricate approach due to their increased power requirements.

1. **Energy Assessment:** The first step in designing an off-grid solar system is to assess the energy needs of the farm or homestead. This involves calculating the total energy consumption of various components, including lighting, heating, cooling, water pumps, and any machinery. It's crucial to have an accurate estimate to avoid under-sizing or over-sizing the system.

2. **Solar Resource Analysis:** The solar resource varies based on geographic location, season, and local weather patterns. Before designing the system, a thorough solar resource analysis is essential to determine the amount of sunlight the location receives throughout the year. This information helps in sizing the solar array appropriately.

3. **Sizing Solar Array and Batteries:** Based on the energy consumption and solar resource analysis, the solar array and battery bank must be sized correctly. Oversizing the solar array can lead to the wastage of resources, while undersizing can result in insufficient energy production. Batteries should have enough capacity to store surplus energy for use during cloudy days or nights.

4. **Inverter and Charge Controller Selection:** Inverters and charge controllers are critical components of off-grid solar systems. The inverter converts DC solar energy into AC electricity for household use. The charge controller regulates the flow of energy from solar panels to the batteries, preventing overcharging. The selection of these components depends on the system's voltage, load requirements, and expected power surges.

5. **Backup Generator Integration:** In regions with prolonged periods of low sunlight, integrating a backup generator powered by a renewable fuel source, such as biogas or biodiesel, can enhance system reliability. The generator can be set to automatically kick in when battery levels drop below a certain threshold.

6. **Wiring and Safety Measures:** Proper wiring and safety measures are crucial to prevent accidents and ensure the longevity of the system. Adequate grounding, surge protection, and adherence to local electrical codes are essential aspects of system design.

Large-Scale Energy Storage Solutions

One of the main challenges in off-grid solar systems for larger applications is effective energy storage. Energy storage solutions play a pivotal role in providing a consistent power supply, especially during periods of low solar irradiance. Here are some notable energy storage options:

1. **Lead-Acid Batteries:** These batteries have been a traditional choice for off-grid systems due to their affordability. However, they have limitations in terms of lifespan and depth of discharge, which can impact the overall system performance.

2. **Lithium-Ion Batteries:** Lithium-ion batteries have gained popularity due to their higher energy density, longer lifespan, and deeper discharge capabilities compared to lead-acid batteries. Although they come at a higher upfront cost, their overall cost-effectiveness over their lifespan is noteworthy.

3. **Flow Batteries:** Flow batteries store energy in liquid electrolytes, allowing for scalable storage capacity. They offer advantages in terms of longer cycle life and faster response times, making them suitable for applications with fluctuating energy demands.

4. **Hydrogen Fuel Cells:** Hydrogen fuel cells provide a unique approach to energy storage. Excess energy is used to produce hydrogen through electrolysis, which can later be used to generate electricity when solar energy is insufficient. While this technology is promising, it currently faces challenges in terms of efficiency and infrastructure.

Managing Power for Multi-Structure Off-Grid Systems

In cases where a single off-grid solar system caters to multiple structures within a farm or homestead, effective power management becomes crucial to ensure equitable distribution of energy. Here are key considerations:

1. **Load Distribution:** Different structures may have varying energy demands. Prioritize essential loads like refrigeration, lighting, and water pumps. Non-essential loads can be scheduled to run when surplus energy is available.

2. **Sub-Metering:** Implement sub-metering to track energy consumption in individual structures. This data helps in identifying energy-intensive areas and optimizing usage patterns.

3. **Smart Energy Management Systems:** Advanced energy management systems can automatically control energy distribution based on real-time data from solar arrays, batteries, and energy consumption. These systems maximize efficiency by intelligently managing power flow.

4. **Energy Monitoring and Maintenance:** Regular monitoring of the system's performance is crucial. Anomalies in energy production or consumption can be detected early, allowing for timely maintenance to prevent system failures.

5. **Community Engagement:** In the context of larger farms or homesteads, involving the community in energy conservation efforts can lead to a more responsible and efficient use of energy resources.

The application of off-grid solar systems for larger setups such as farms and homesteads necessitates meticulous planning, accurate energy assessment, appropriate technology selection, and efficient power management. These systems not only provide a sustainable energy solution but also empower remote communities to have a reliable and self-sufficient power source. As technology continues to advance, off-grid solar systems will likely become even more accessible, efficient, and essential for addressing energy challenges in remote areas.

CHAPTER 3: OFF-GRID SOLAR ENTREPRENEURSHIP AND COMMUNITY PROJECTS

In an era defined by environmental concerns and the growing need for sustainable energy sources, the spotlight has turned decisively toward solar power. The potency of harnessing the sun's energy as a clean and renewable source has spurred innovation, birthing a plethora of opportunities for both entrepreneurs and community-driven initiatives. This chapter delves into the dynamic landscape of off-grid solar entrepreneurship and its intersection with community projects. It explores the nuances of starting a solar installation business, the impact of community-based solar initiatives and cooperative projects, and the role of solar empowerment within non-profit ventures.

Starting a Solar Installation Business

The solar energy industry stands as a beacon of hope in the global quest for sustainable energy solutions. Entrepreneurs with a passion for clean energy and a penchant for innovation have found themselves presented with a unique avenue: starting a solar installation business. This endeavor offers not only the promise of financial success but also the opportunity to contribute meaningfully to a greener future.

Embarking on this journey requires a comprehensive understanding of the solar energy landscape. Potential business owners must delve into the technical aspects of solar panels, inverters, batteries, and other essential components of a solar system. Moreover, an in-depth knowledge of local regulations, incentives, and permits is crucial to navigating the often complex bureaucratic environment surrounding renewable energy projects.

A successful solar installation business hinges on several key factors:

1. **Expertise and Quality**: In a rapidly evolving field, staying updated with the latest technological advancements is vital. Providing top-notch installation services and using high-quality components ensures customer satisfaction and system longevity.
2. **Customization**: Every client's energy needs and infrastructure are unique. Tailoring solar solutions to individual requirements showcases a commitment to effective energy utilization.
3. **Financial Savviness**: Demonstrating the financial benefits of transitioning to solar power, including potential savings on utility bills and available incentives, can be a compelling selling point.
4. **Networking**: Building relationships within the local solar community, including suppliers and potential partners, can open doors to collaborations and insights.
5. **Marketing**: Effectively conveying the environmental and economic advantages of solar energy through strategic marketing efforts can attract a wider customer base.

Community-Based Solar Initiatives and Cooperative Projects

In the realm of solar energy, where the power of the sun unites with the spirit of collaboration, community-based solar initiatives and cooperative projects emerge as beacons of collective progress. These endeavors redefine how energy is harnessed, shared, and celebrated, transforming the way communities engage with renewable resources and shaping a future marked by shared responsibility and sustainable growth.

Empowering the Collective

Community-based solar initiatives are a testament to the principle that energy is not just an individual pursuit, but a collective endeavor. These projects bring together community members, pooling resources and knowledge to establish solar power systems that benefit multiple stakeholders. By leveraging economies of scale and shared investments, these initiatives make solar energy accessible to a broader demographic, including those who may not have the means to install solar panels on their individual properties.

The Cooperative Tapestry

Cooperative solar projects weave a cooperative tapestry where multiple participants collaborate to establish and maintain a shared solar installation. These endeavors may take the form of community solar gardens, where arrays of panels are installed on a designated piece of land, or they could involve installing solar panels on communal buildings like schools, libraries, or community centers. The beauty of these projects lies in their inclusive nature, transcending socio-economic barriers and fostering a sense of ownership among participants.

Dividends Beyond Energy

The impact of community-based solar initiatives extends beyond the generation of clean energy. These projects foster a sense of pride and unity within the community, as members work together to shape a sustainable future. They serve as educational platforms, providing opportunities for learning about solar technology, energy conservation, and the interconnectedness of environmental and social well-being.

Shared Costs, Shared Benefits

One of the compelling aspects of these initiatives is the distribution of costs and benefits. Participants share the initial investment and ongoing maintenance expenses, allowing for a more equitable distribution of financial burdens. As the solar panels generate energy, participants reap the benefits in the form of reduced electricity bills, translating into tangible savings that directly impact their lives.

A Blueprint for Sustainability

Community-based solar initiatives and cooperative projects serve as blueprints for sustainability that extend beyond energy. They foster a sense of ownership and stewardship among participants, nurturing a culture of responsibility towards the environment. By reducing reliance on fossil fuels and decreasing carbon footprints, these projects contribute to the global effort to mitigate climate change while demonstrating the power of unity in achieving shared goals.

Weaving a Sustainable Future

In the symphony of renewable energy, community-based solar initiatives and cooperative projects compose a harmonious refrain that resonates with hope and possibility. These endeavors echo the spirit of togetherness, demonstrating that through collaboration, communities can shape their destiny and weave a sustainable future. As solar panels capture the sun's energy, they also capture the collective aspiration for a world where progress is defined not only by technological advancement, but by the strength of community bonds and a commitment to safeguarding the planet for generations to come.

CHAPTER 4: FUTURE TRENDS AND INNOVATIONS IN OFF-GRID SOLAR POWER

In an era where sustainable and clean energy solutions have become paramount, the evolution of off-grid solar power has emerged as a remarkable force driving change. As the global community seeks alternatives to traditional fossil fuels, the realm of solar energy stands at the forefront of innovation.

Advancements in Solar Panel Technology

Solar panels, also known as photovoltaic (PV) panels, are the heart of any solar power system. They convert sunlight into electricity through the photovoltaic effect, providing a clean and sustainable source of energy. Over the years, significant advancements have been made in solar panel technology, resulting in improved efficiency, affordability, and versatility.

One of the most notable trends in solar panel technology is the constant drive to enhance efficiency. Early solar panels had relatively low-efficiency rates, converting only a small portion of sunlight into usable electricity. However, ongoing research and development have led to the creation of highly efficient panels that can convert a larger percentage of sunlight into energy. This increased efficiency translates to greater electricity generation for a given surface area, making solar power a more viable option for a range of applications.

Traditionally, solar panels were rigid and bulky, limiting their applications to rooftops and open fields. However, the development of thin-film solar panels has revolutionized the industry. Thin-film panels are lightweight and flexible, allowing them to be integrated into a variety of surfaces, including curved structures and even clothing. This innovation opens up new possibilities for off-grid solar power in urban environments and portable electronics.

Tandem solar cells represent another breakthrough in solar panel technology. These cells stack multiple layers of solar materials with different absorption properties, effectively capturing a broader spectrum of sunlight. By utilizing complementary materials, tandem cells can achieve higher efficiency levels than traditional single-layer cells. This advancement has the potential to significantly boost energy production, particularly in regions with varying sunlight conditions.

Perovskite solar cells have garnered immense attention due to their low production cost and impressive efficiency gains. These cells use perovskite materials, often a hybrid organic-inorganic lead or tin halide-based compound, to absorb sunlight and generate electricity. While still undergoing refinement to address durability and stability concerns, perovskite solar cells hold promise for driving down the cost of solar power and expanding its accessibility.

Energy Storage Breakthroughs and Emerging Battery Technologies

While solar panels excel at generating electricity during daylight hours, effective energy storage systems are essential to provide power during nighttime and cloudy periods. The evolution of off-grid solar power hinges not only on efficient energy generation but also on robust energy storage solutions. Recent breakthroughs in battery technologies have paved the way for enhanced energy storage capabilities.

Lithium-ion batteries have become the de facto standard for energy storage in a multitude of applications, from smartphones to electric vehicles. These batteries offer high energy density, fast charging capabilities, and a relatively long lifespan. In the context of off-grid solar power, lithium-ion batteries provide a reliable solution for storing excess energy generated during sunny periods for use during less favorable conditions.

Solid-state batteries represent the next frontier in energy storage technology. Unlike traditional lithium-ion batteries, which use liquid electrolytes, solid-state batteries employ solid electrolytes. This design not only enhances safety by reducing the risk of leakage or thermal runaway but also offers the potential for higher energy densities and faster charging rates. The integration of solid-state batteries into off-grid solar power systems could revolutionize energy storage efficiency.

Flow batteries offer a unique approach to energy storage by utilizing two separate electrolyte solutions that flow through an electrochemical cell. This design allows for decoupling the energy storage capacity from the power output, enabling scalability and longer-duration energy storage. Flow batteries are particularly well-suited for large-scale off-grid applications, such as microgrids and remote communities.

The Evolution of Off-Grid Solar Power Systems

The integration of advanced solar panel technology and innovative energy storage solutions has ushered in a new era of off-grid solar power systems. These systems are no longer confined to remote cabins or research projects; they are becoming viable alternatives to traditional grid-connected setups in various settings.

Off-grid solar power systems are increasingly finding their way into residential settings. Homeowners are leveraging solar panels and energy storage solutions to become more self-reliant and reduce their dependence on traditional utility grids. This trend is particularly relevant in regions prone to power outages or those with limited access to reliable electricity sources.

In remote and underserved areas, off-grid solar power has emerged as a lifeline. Traditional grid infrastructure can be prohibitively expensive to establish in such regions. Off-grid solar power systems provide a cost-effective and sustainable solution for bringing electricity to schools, healthcare facilities, and communities that have long been in the dark.

Natural disasters and humanitarian crises often disrupt essential services, including electricity supply. Off-grid solar power systems can be rapidly deployed to provide emergency energy sources for charging communication devices, running medical equipment, and maintaining basic living conditions in affected areas. Their portability and ease of installation make them indispensable tools in disaster relief efforts.

Industries and businesses are also recognizing the benefits of off-grid solar power. In remote mining operations, for instance, solar power can reduce reliance on costly diesel generators and decrease the environmental impact. Similarly, businesses seeking to align with sustainable practices can integrate off-grid solar power systems to power their operations while reducing their carbon footprint.

CONCLUSION

From the very inception of this book, my aim was to equip you with the knowledge, skills, and inspiration to embark on a transformative path—one that leads to energy independence, environmental stewardship, and a lifestyle aligned with nature.

Throughout these pages, we've explored the intricacies of off-grid solar power, transcending the boundaries of traditional energy sources and embracing the boundless potential of the sun. We've delved into the mechanics of solar panel installation, the art of energy management, and the nuances of crafting a sustainable existence. We've examined the diversity of applications, from tiny homes to cabins, RVs to boats, uncovering how off-grid solar power seamlessly integrates with various lifestyles.

But this journey doesn't end here. In fact, this is just the beginning—a prelude to a life empowered by the sun's rays and driven by a commitment to preserving our planet. As you venture into the world armed with the insights gleaned from these pages, I urge you to remember that you are not merely a consumer of energy; you are a creator of change. Each time you harness the power of the sun, you contribute to a cleaner, greener future.

Off-grid solar power isn't just about technicalities; it's a mindset, a lifestyle, and a philosophy. It's a statement that you value self-reliance, that you're attuned to the rhythms of the earth, and that you're ready to embrace the challenges and rewards of conscious living. Whether you're basking in the warmth of solar-heated water, relishing the glow of LED lights powered by the sun, or marveling at the way your energy-efficient home blends seamlessly with nature, know that you are part of a movement that's shaping a better world for generations to come.

The journey of off-grid solar power is dynamic, just like the sun that fuels it. It's a journey of continuous learning, innovation, and growth. As technology evolves and new discoveries unfold, I encourage you to stay curious, stay informed, and keep pushing the boundaries of what's possible. And always remember, the sun is an unwavering ally—shining down on you, ready to be harnessed into a force that propels you toward a brighter, cleaner, and more sustainable future.

Thank you for embarking on this adventure with me. It has been an honor and a privilege to be your guide on this odyssey. As you step into the world armed with the wisdom and confidence gained from *The DIY Off-Grid Solar*

Power Bible, I am confident that you will not only transform your own life but also inspire others to follow suit. Together, we are rewriting the narrative of energy consumption and embracing a future that's bright in every sense of the word.

Here's to the power of the sun, the power within you, and the power to create a world that thrives on sustainable, off-grid solar energy.

Made in the USA
Las Vegas, NV
20 April 2024

88841971R00109